Studies in Computational Intelligence

Volume 508

Series Editor

J. Kacprzyk, Warsaw, Poland

For further volumes:
http://www.springer.com/series/7092

Mohd. Samar Ansari

Non-Linear Feedback Neural Networks

VLSI Implementations and Applications

 Springer

Mohd. Samar Ansari
Electronics Engineering
Aligarh Muslim University
Aligarh, UP
India

ISSN 1860-949X   ISSN 1860-9503 (electronic)
ISBN 978-81-322-2896-7   ISBN 978-81-322-1563-9 (eBook)
DOI 10.1007/978-81-322-1563-9
Springer New Delhi Heidelberg New York Dordrecht London

Printed on acid-free paper

Springer is part of Springer Science+Business Media (www.springer.com)

It is said that a good marriage is the closest thing to heaven on earth. In that light, I dedicate this book to the person who has facilitated that glimpse of paradise for me ...

To

My Wife Asra

With Love

Preface

Artificial Neural Networks (ANNs), having a highly parallel architecture, have emerged as a new paradigm for solving computationally intensive tasks using collective computation in a network of neurons. They can be considered as analog computers relying on simplified models of neurons. The essential difference between the ANN's distributed computation and a modern day digital computer's functioning is that in the case of a digital computer, the processing is often centralized at one location by means of a single processor (aptly referred to as the 'uniprocessor' architecture), whereas the collective computation model of a neural network depends upon the simultaneous working of hundreds of thousands of analog neural processors. Each of these 'neurons' have very limited computational power when considered as a separate entity. However, the real power comes with their acting in unison. The immense computational abilities of a neural network are a perfect example of the benefits of teamwork!

Recurrent neural networks, which are essentially ANNs employing feedback interconnections between the neurons, were extensively explored after the concept of Lyapunov (or 'energy') functions, as a means of understanding the complex dynamics of such networks, was introduced by Hopfield and Tank. Their architecture, called the Hopfield Network, was implementable in hardware, and although it became very popular, many limitations like convergence to infeasible solutions and the requirement of a large number of neurons and interconnection weights were revealed when attempts were made to apply it to practical applications. These drawbacks warranted the exploration of alternative neural network architectures which are amenable to hardware realizations. The Nonlinear Synapse Neural Network (NOSYNN) has been proposed as one such alternative which alleviates the problems that plagued the Hopfield Network.

This book deals with VLSI implementations and applications of the NOSYNN type of nonlinear feedback neural networks. These networks have been shown to be better performing than their Hopfield Neural Network (HNN)-based counterparts, in the sense that their convergence to the exact solution is fast and guaranteed. This improvement in the performance is due to underlying difference in the nature of feedback between the HNN and the NOSYNN. While the HNN employs linear feedback (typically implemented using resistors), the NOSYNN employs nonlinear feedback (typically implemented using voltage-mode comparators). This difference in hardware also carries over to a difference in the dynamical properties of the two networks and makes the energy functions of the two networks vastly different. While the HNN has a quadratic form of the energy function, the NO-SYNN has transcendental terms in its energy function which account for better and faster convergence characteristics.

In this text, the NOSYNN architecture has been chosen as the starting point of the exploration for better hardware implementations of ANNs. Thereafter, the content progresses in two dimensions. First, the voltage-mode NOSYNN has been reconfigured and applied to a new problem, viz. the solution of linear equations. As has been mentioned above, the NOSYNN-based neural network for solving linear equations has an associated energy function which contains transcendental terms as opposed to the quadratic terms in the energy functions for the Hopfield network and its variants. It is shown that the network has only a unique minimum in the energy landscape, and moreover, the global minimum coincides exactly with the solution point of the system of linear equations. Thereafter, it has been shown that two other important problems of mathematical programming: linear programming problem (LPP) and quadratic programming problem (QPP) could also be solved by incorporating small modifications in the voltage-mode network proposed for the solution of linear equations.

Second, a 'mixed'-mode (MM) implementation for the NOSYNN has been discussed. Applications of the 'mixed'-mode neural circuit in solving linear equations, LPP, QPP, graph coloring, and ranking of numbers, are explained in detail. In the so-called 'mixed'-mode hardware realization, neuron states are represented as voltages whereas the synaptic interconnections convey information in the form of currents. It has been shown that the mixed-mode implementation of the NOSYNN leads to reduction in the overall circuit complexity, as compared to the voltage-mode realization, by eliminating the resistors employed as synaptic weights. Two different VLSI realizations of the 'mixed'-mode networks are discussed. The first employs Differential Voltage Current Conveyors (DVCCs) to implement voltage comparators with current outputs. The second class of realizations use Operational Transconductance Amplifiers (OTAs) to provide the required voltage comparison at the inputs of the comparators.

Why this Book?

A very pertinent question that would come to the mind of a person coming across this book, while searching for a book on feedback neural networks, is 'What does this text has to offer which any other book does not?' The answer to the question lies in the nature of content. While a multitude of (very good) books on neural networks are heavy on theory and the related mathematics without dwelling on the actual hardware implementations, this particular text focusses on the intricacies involved when a neural circuit is actually realized in hardware. This book does not intend to replace the already established books on the subject. Rather, it offers the readers an additional cue about how to actually port the neural circuit, leaving behind all the mathematics, to a real-world circuit, for instance, a CMOS realization. The block diagram representations that abound in many of the existing texts are sufficient for developing an understanding about the intended working of the network. However, such a representation generally assumes ideal values and behavior of the various components, which is seldom the case in an actual real-world implementation. This has led to a severe dent on the neural network field in general, with critics saying that neural networks promise the moon but deliver nothing. Almost entirety of such statements are issued by researchers who are attracted by the field only to find it full of perfectly working mathematical models which turn into nonperforming entities in a breadboard implementation. It is the contention of the author that a significant number of such misplaced ideas about neural networks can be eliminated if interested persons are provided the required help in actual hardware design and testing of such circuits.

Although it is true that the real power of neural networks lies in massively parallel structures, containing hundreds of neurons, capable of solving combinatorial (and other) problems comprising a large number of variables; the fact should not prevent a book from being able to provide small-sized scaled versions of those huge networks, just to make a reader familiar with the actual operation of the network. The approach followed in this book is simply to start with the most simple case, understand its operation, get its maths right, test it in hardware, and then move on to somewhat bigger problems. For instance, for every linear equation solver discussed in the book, first a small circuit capable of solving just two linear equations is presented and explained. It is the belief of the author that the concept of energy function is easily grasped for networks with low neuron counts. Once the operation, maths and the hardware of the two variable linear equation solver is complete, the text moves on to slightly more complex circuits before finally dealing with the generalized version of the network.

Prerequisites

For a good understanding of the topics covered in this book, a reader should have knowledge about electronic amplifiers, particularly the operational amplifier. A basic knowledge of circuit theory is assumed. On the mathematics front, the reader should have studied differential and integral calculus as well as mathematical optimisation.

Acknowledgments

Writing a book is seldom a 'one-man' effort. This remains true for this book as well. There are many whom I would like to thank and acknowledge.

I am at loss to find words suitable enough to express my gratitude to my parents for their unfaltering love and support. I am indebted to them for the difficulties, and loneliness, that they endure while I am away from them. I also wish to thank my younger brother Sarim for his deep concern and support in my work.

Thanks are also due to my wife, Asra, for her patience and faith in me. My dual responsibility of work and family has been greatly lessened thanks to her understanding nature.

I take this opportunity to thank Dr. Syed Atiqur Rahman for introducing me to the nonlinear feedback architecture in neural networks. It is through his concern and continuous help that I have been able to write this book. I am also thankful to Prof. Ekram Khan, Dr. Omar Farooq, and Dr. Sudhanshu Maheshwari for their time, encouragement, and helpful suggestions.

I also thank my friends and colleagues Parveen Beg, Mohd. Sharique, Syed Javed Arif, Ale Imran, Naushad Alam, and Beenish Zia for the different ways in which they have helped me.

Since the present document was prepared in LATEX, thanks are also due to the thousands of individuals who have coded for the LATEX project for free. It is due to their efforts that we can generate professionally typeset PDFs now.

Acknowledgments

Contents

Figures

Tables

Chapter 1
Introduction

1.1 Neural Networks

A digital computer can outperform a human brain in performing long calculations and solving complex mathematical problems. However, when it comes to tasks other than number crunching, the human brain possesses several advantages over a digital computer. A person can easily recognize a face in a crowd, even when the lighting is poor and only a side (or partial) view is available. Similarly, speech recognition is another task that all people can do easily. A human being can understand the half-formed words of a baby, a drawl, and different accents. Even after decades of intense research, digital computers still lag behind at this level. The brain is also remarkably robust in the sense that the death of a few cells does not cause it to stop functioning. Compared to this, computers do not, in general, survive degradation in the Central Processing Unit (CPU) [1].

A lot of research efforts have been put into trying to emulate the computational capabilities exhibited by the human brain. Although the precise working mechanism of the brain still remains a mystery, it is known that the computations are done by a highly interconnected network of neurons, which communicate with each other by sending electrical pulses through the neural wiring consisting of axons, synapses and dendrites. Taking cue from this high performance biological neural network, artificial neural networks composed of interconnected artificial neurons were developed. The artificial neurons could either be implemented in software as programming constructs or in hardware as electronic models mimicking the functionality of the biological neuron.

The earliest known computational model of the neuron is attributed to McCulloch and Pitts [2], and depicts the neuron as a switch that receives inputs from other neurons and, depending on the total weighted input, is either activated or remains inactive [2]. The weight, by which an input from another cell is multiplied, corresponds to the strength of a synapse—the neural contact between the nerve cells. These weights can be both positive (excitatory) or negative (inhibitory).

M. S. Ansari, *Non-Linear Feedback Neural Networks*,
Studies in Computational Intelligence 508, DOI: 10.1007/978-81-322-1563-9_1,
© Springer India 2014

The next major development in neural network technology arrived in 1949 with a book "The Organization of Behavior" written by Donald Hebb in which a hypothesis of learning based on the mechanism of neural plasticity, now known as Hebbian learning, was presented [3]. In 1958, Rosenblatt demonstrated that simple networks of neurons, called Perceptrons, could learn from examples. The Perceptron, although built using primitive hardware, was the first practical artificial neural network [4].

In their book titled "Perceptrons", Minsky and Papert pointed out two key issues with the computational machines that processed neural networks [5]. Firstly, Rosenblatt's perceptrons could not perform the calculation of parity and therefore were unable to learn to evaluate the logical function of Exclusive-OR(XOR). Secondly, Minsky and Papert argued that a network comprised of two (or more) layers of perceptrons might not exhibit the properties which are similar to the basic perceptron. The results of Minsky & Papert's analysis led them to the conclusion that, despite the fact that perceptrons were 'interesting' to study, ultimately perceptrons and their extensions were a 'sterile' direction of research [6].

> There is no reason to suppose that any of these virtues carry over to the many layered version. Nevertheless, we consider it to be an important research problem to elucidate (or reject) our intuitive judgement that the extension is sterile (p. 231, Perceptrons [5]).

Between 1959 and 1960, Widrow and Hoff developed the ADALINE (Adaptive Linear Elements) and MADALINE (Multiple Adaptive Linear Elements) models [7]. These were the first artificial neural networks that could be applied to real problems. The ADALINE model is used as a filter to remove echoes from telephone lines. The capabilities of both these models were again proved limited by Minsky and Papert (1969) [5]. In 1975, Werbos came up with the Back Propagation algorithm which effectively solved the Exclusive-OR implementation problem that plagued the Perceptron [8]. In the same year, Fukushima presented a multi-layered neural network and called it the Cognitron [9].

In his seminal paper of 1982, Hopfield put forward the idea that the approach to artificial intelligence should not be to purely imitate the human brain but instead to use the basic concepts to build machines that could solve dynamic problems [10]. His ideas gave birth to a new class of neural networks that over time came to be known as Hopfield neural Networks (HNNs). In 1986, Rumelhart, Hinton and Williams reported on the developments of the Back Propagation algorithm discussing how back propagation learning had emerged as the most popular learning method for the training of multi-layer perceptrons [11].

Other significant milestones in the development of neural network architectures include the advent of Radial Basis Function Neural Networks [12], Self Organizing Maps [13], Modular Neural Networks [14], Cellular Neural Networks [15], Support Vector Machines [16], and unsupervised learning procedures using deep learning algorithms [17].

1.2 Applications of Neural Networks

As mentioned in the previous section, networks of interconnected neurons may either be implemented in software or as hard-wired structures. Therefore, it is necessary that a distinction is made between the two. From this point onwards, the term Artificial Neural Network (ANN) is intended to imply a software-based or simulated structure—an algorithm rather than a physical system. Although slow in comparison, ANNs are much cheaper to design and implement than their solid-state counterparts [18], which will, from now on, be referred to as Neural Networks (NNs) throughout this book.

ANNs have been implemented to solve a variety of problems including, but not limited to, speech recognition [19], handwritten character recognition [20], fingerprint recognition [21], underwater sonar identification [22], meteorological classification [23], automatic vehicle control [24], diagnosis of hypertension [25], detection of heart abnormalities [26], detection of explosives [27], prediction of bank failure [28], prediction of stock market performance [29], staff scheduling [30], retail inventories optimization [31], signal processing (neural filtering) [32], routing optimization [33], electrical load estimation [34], data mining (time series analysis) [35], and email spam filtering [36].

Although the list of existing (and potential) applications of ANNs is impressive, these software implementations are, in general, sequential in nature and therefore, are not suitable for applications requiring real-time data processing (for example, streaming video compression) [37]. Further, running of an algorithm may also require an operating system which may not be feasible on a portable system where a dedicated, compact, low-power solution is desirable. Specialized neural network hardware offers the following advantages [38]:

- Speed: Dedicated hardware can offer very high computational power in the neural domain where parallelism and distributed computing are inherently present.
- Cost: A hardware implementation can provide margins for reducing the total cost of the system by lowering the component count and decreasing power requirement.
- Graceful Degradation: The lack of redundancy in uni-processor based systems makes applications running on such systems vulnerable to faults in the system (fail-stop operations). In contrast to this, hardware neural networks, by virtue of their distributed and parallel architecture, allow functionality albeit with reduced performance even if components become faulty.

Although not as widespread as ANNs, NNs do find their place in real world applications. Examples include applications in high energy physics [39], pattern recognition [40], image/object recognition [41], generic image/video processing [42], direct feedback control [43], real-time embedded control [44], audio synthesis [45], and optical character/handwriting recognition [46].

1.3 Hardware for Neural Networks

The applications of NNs mentioned in the previous section consist of a mix of analog and digital hardware neural networks. Although there is a perceived competition between 'analog' and 'digital' NNs, this competition is a misconception. Both these implementations have their own advantages and limitations, and the selection of any one design methodology would depend on a number of factors like speed, cost, accuracy, etc.

An analog implementation is usually efficient in terms of chip area and processing speed, but this comes at the price of a limited accuracy of the network components. In a digital implementation, on the other hand, accuracy is achieved at the cost of efficiency (e.g. relatively larger chip area, higher cost, and more power consumption) [38]. This amounts to a trade-off between the accuracy of the implementation and the efficiency of its performance. In a digital neuron, synaptic weights are stored in latches, shift registers or memories. Adders, subtractors and multipliers are available as standard circuits, and non-linear activation functions can be implemented using look-up tables or using adders, multipliers, etc. [38]. A digital implementation enjoys advantages like simplicity, high signal-to-noise ratio, easily achievable cascadability and flexibility, along with some demerits like slower operation.

In an analog neuron, weights are usually stored using one of the following: resistors, charge-coupled devices, capacitors, and floating-gate electrically erasable programmable read-only memories (EEPROMs). In very large scale integration (VLSI) realizations, a variable resistor acting as a weight can be implemented as a circuit employing two MOSFETs. Scalar product and non-linear mapping actions are performed by a summing amplifier with saturation. Analog elements are generally smaller and simpler than their digital counterparts [38]. Furthermore, as the real world is 'analog' it would seem prudent to have NNs working in the analog domain. Also, since it has been shown time and again that analog computation is faster than conventional digital methods [47, 48], analog NNs are preferable over digital NNs for compact, energy efficient and fast applications. However, obtaining consistently precise analog circuits especially to compensate for variations in temperature and control voltages demands sophisticated design and fabrication.

Some of the early fully developed analog NN chips include Intel's Electrically Tunable Analog Neural Network (ETANN) 80170NX [49] and Synaptic's Silicon Retina [50]. The ETANN is a general purpose neurochip having 64 fully connected neurons and 10,240 programmable synapses. The Mod2 Neurocomputer employs 12 such ETANN chips for real-time image processing [51]. A CMOS feed-forward chip with on-chip error reduction hardware for real-time adaptation has also been developed [52]. An analog synapse model using MOSFETs in standard 0.35 μm CMOS fabrication process has also been proposed [53]. This synapse has been employed in a VLSI architecture consisting of 2,176 synapses for the purpose of fingerprint feature extraction [53]. Implementation of a signal processing circuit for a feedback neural network using sub-threshold analog VLSI in mixed-mode approach, where state variables are represented by voltages while neural signals are conveyed as currents, has also been reported [54].

There have also been attempts at designing digitally programmable analog building blocks for NN implementations. Hybrid neurochips combine digital and analog technologies while attempting to get the best features of both. The internal processing is normally analog with weights being set digitally. One example is the hybrid Neuro-Classifier which has 70 analog inputs, 6 hidden nodes, and one analog output with 5-bit digital weights [55]. A hardware efficient matrix-vector multiplier architecture for NNs with digitally stored synapse strengths has also been developed [56].

Out of the many neural network architectures which have been implemented in hardware, the Hopfield neural network (HNN) has attracted the most attention from the neural networks community as a model of analog circuit computation [10]. It is essentially a feedback neural network having non-linear neurons with linear feedback, the details of which are presented in the next chapter. Many other models that appeared in the technical literature after the advent of Hopfield networks, like the bidirectional associative memories [57], Botzmann machines [58], Q-state attractor neural networks [59], etc., are either direct variants or generalizations of the Hopfield network [60]. Typically, the convergence of a dynamical system is explained with the help of an associated Lyapunov function (commonly referred to as the energy function). For the Hopfield network, the energy function contains quadratic terms [10]. The HNN dynamics is such that it converges to a local minimum of its energy function; therefore, the HNN can only solve unconstrained minimization tasks. However, most HNN applications solve constrained optimisation problems by modifying the energy function to incorporate terms that penalize infeasible configurations. Each of these 'constraint terms' is associated with a weight. Choosing the weights is often a matter of trial and error [61]. Moreover, attempts to use HNNs in practical applications have revealed many difficulties. These include convergence to infeasible solutions and the requirement of a large number of neurons and interconnection weights [62, 63]. These drawbacks led to the exploration of alternative neural network architectures.

The Nonlinear Synapse Neural Network (NOSYNN) is one such neural network architecture[63]. The use of nonlinear synapse leads to a NOSYNN energy function that involves transcendental functions, which is fundamentally different from the earlier quadratic form of energy function associated with Hopfield Network and its variations. It has been shown that use of the NOSYNN architecture can alleviate the problem of convergence to spurious states inherent in the Hopfield Network [63]. Also, NOSYNN-based solutions generally require much lesser number of neurons and interconnection weights than Hopfield network based approaches [63]. Since its introduction, the NOSYNN has been effectively employed for the solution of computational problems like graph colouring [64], and ranking (sorting) of numbers [65]. Considering the advantages offered by the NOSYNN over the standard Hopfield network (and its variants), the NOSYNN has been selected as the architecture of choice for the various applications discussed in this text.

1.4 Outline of Contents

It has been discussed in Sect. 1.2 that in order to obtain the actual benefits of neural computation, like higher speed and lower cost, it is imperative that the networks are realized in hardware. This fact forms the motivation for the work contained in this book. Satisfactory and efficient hardware implementation of neural networks is still an open area of research and many design challenges lie ahead in this field [66–70]. The NOSYNN architecture presented in the previous section has been chosen as the starting point of the exploration for better hardware implementations. Thereafter, the content may be roughly classified in two distinct categories.

Firstly, the voltage-mode NOSYNN has been reconfigured and applied to a new set of problems *viz.* the solution of linear equations and linear and quadratic programming problems. The selection of NOSYNN as the architecture of choice is justified by highlighting the shortcomings associated with the use of another popular architecture *viz.* the Hopfield neural network. It is shown that the standard Hopfield network cannot be used for solving linear equations and suitable modifications are discussed. It is further shown that even the modified Hopfield network based linear equation solver has limited applicability thereby necessitating the need for an alternative network architecture. The NOSYNN architecture is selected as the alternative to Hopfield Network. To solve a system of simultaneous linear equations, the 'energy' function associated with the NOSYNN has been modified suitably so as to yield a global minima that coincides exactly with the solution point of the set of equations. Thereafter, it has been shown that two other important problems of mathematical programming: linear programming problem (LPP) and quadratic programming problem (QPP) could also be solved by incorporating small modifications in the voltage-mode network proposed for the solution of linear equations. Prior to considering new problems, applications for which the NOSYNN has already been employed successfully, i.e. the Graph Colouring Problem [63], and ranking of numbers [65], are discussed. It is shown that the quality of the obtained colouring results can be improved by a simple alteration in the circuit.

Secondly, a relatively new implementation for the NOSYNN is discussed. This is the 'mixed'-mode (MM) non-linear feedback neural network. In the so called 'mixed'-mode hardware realization, neuron states are represented as voltages whereas the synaptic interconnections convey information in the form of currents. It has been shown that the mixed-mode implementation of the NOSYNN leads to reduction in the overall circuit complexity, as compared to the existing voltage-mode realization, by eliminating the resistors employed as synaptic weights. Alternative mixed-mode circuit realizations are presented for all the applications considered in the book.

1.5 Organization of the Chapters

The content is organized as follows.

Chapter 2 contains the background information relevant to the work presented in later chapters. Technical details of both the Hopfield neural network and the NOSYNN are presented. Applications of the NOSYNN architecture to graph colouring and sorting problems are also discussed wherein it is shown that a slight alteration in an existing NOSYNN based circuit for graph colouring can result in improvements in the performance. Thereafter, the different problems for which hardware solutions are presented in this book are explained in detail. This is followed by a review of the existing neural network based approaches pertinent to the chosen problems.

Chapter 3 starts with an exploration into the possibility of applying Hopfield neural network based approaches for the solution of simultaneous linear equations. It is shown that Hopfield's original network is not suitable for the task of solving linear equations. Thereafter, modifications that need to be incorporated into the standard Hopfield network in order to make the network amenable for the chosen application are discussed. Furthermore, it is shown that even the modified Hopfield network has limited applicability when applied for solving linear equations thereby necessitating the need of an alternative architecture. Thereafter, a voltage-mode non-linear feedback neural circuit for solving linear equations is discussed. The NOSYNN architecture is employed thereby causing the introduction of transcendental terms in the energy function associated with the proposed network. The proof of the energy function has been given and it is shown that the gradient network converges exactly to the solution of the system of equations. The operation of the circuit is verified by simulations done using the PSPICE program. Moreover, hardware implementation results for small-sized problems are presented which further confirm the circuit operation.

Chapter 4 contains details of a mixed-mode hardware implementation of the linear equation solver of Chap. 3. The state variables are represented by voltages while neural signals are conveyed as currents thereby causing the elimination of the passive resistors connected in the synaptic feedback paths. A relatively new building block for analog signal processing, the digitally-controlled differential voltage current conveyor (DC-DVCC) is reviewed and is utilized for the non-linear feedback interconnections between neurons. PSPICE simulation results are presented for linear systems of equations of various sizes and are found to be in close agreement with the algebraic solution. Also, the use of CMOS DC-DVCCs and operational amplifiers makes the circuit more suitable for monolithic integration as compared to the voltage-mode network of Chap. 3 where a large number of passive resistors are employed.

In Chap. 5, techniques for the extension of the voltage-mode networks for solving linear equations to solve mathematical programming problems like LPP and QPP are discussed. It is shown that both LPP and QPP can be solved by properly incorporating small modifications in the basic network of the linear equation solver. PSPICE simulations for various sample problems are performed to validate the approach.

Chapter 6 presents Operational Transconductance Amplifier (OTA) based implementations of mixed-mode networks for solving linear equations, graph colouring, ranking of numbers, linear and quadratic programming.

Chapter 7 contains concluding remarks. The chapter also indicates avenues for further study and aspects of the content that motivate further investigation.

References

1. Krogh, A.: What are artificial neural networks? Nat. Biotechnol. **26**(2), 195–197 (2008)
2. McCulloch, W.S., Pitts, W.: A logical calculus of the ideas immanent in nervous activity. Bull. Math. Biol. **5**(4), 115–133 (1943)
3. Hebb, D.O.: The organization of behavior. Wiley, New York (1949)
4. Rosenblatt, F.: The perceptron: a probabilistic model for information storage and organization in the brain. Psychol. Rev. **65**(6), 386–408 (1958)
5. Minsky, M.L., Papert, S.: Perceptrons. MIT Press, Oxford (1969)
6. Pollack, J.B.: No harm intended: Marvin, L. Minsky and Seymour A. Papert. Perceptrons: An Introduction to Computational Geometry, Expanded edition. Cambridge, MA: MIT Press, 1988. p. 292. $12.50 (paper). J. Math. Psychol. **33**(3), 358–365 (1989)
7. Widrow, B., Hoff Jr, M.E.: IRE western electric show and convention. Record **4**(96–104), (1960)
8. Werbos, P.J.: Beyond regression: new tools for prediction and analysis in the behavioral sciences. Ph.D. thesis, Harvard University, Cambridge (1975)
9. Fukushima, K.: Cognitron: a self-organizing multilayered neural network. Biol. Cybern. **2**(3), 121–136 (1975)
10. Hopfield, J.J.: Neural networks and physical systems with emergent collective computational abilities. Proc. Nat. Acad. Sci. **79**(8), 2554–2558 (1982)
11. Rumelhart, D.E., Hinton, G.E., Williams, R.J.: Learning internal representations by error propagation. MIT Press, Cambridge (1986)
12. Moody, J., Darken, C.J.: Fast learning in networks of locally tuned processing units. Neural Comput. **1**(2), 281–294 (1989)
13. Kohonen, T.: Self-organized formation of topologically correct feature maps. Biol. Cybern. **43**(1), 59–69 (1982)
14. Happel, B., Murrey, J.: The design and evolution of modular neural network architectures. Neural Netw. **7**(6), 985–1004 (1994)
15. Manganaro, G., Fortuna, L., Arena, P.: Cellular Neural Networks, 1st edn. Springer, Secaucus (1999)
16. Cortes, C., Vapnik, V.: Support-vector networks. Mach. Learn. **20**, 273–297 (1995)
17. Hinton, G.E., Salakhutdinov, R.R.: Reducing the dimensionality of data with neural networks. Science **313**(5786), 504–507 (2006)
18. Schalkoff, R.J.: Artificial Neural Networks. McGraw-Hill, New York (1997)
19. Sejnowski, T.J., Goldstein, M.H., Jr., Jenkins, R.E., Yuhas, B.P.: Combining visual and acoustic speech signals with a neural network improves intelligibility. In: Advances in Neural Information Processing Systems, pp. 232–239. Morgan Kaufmann Publishers Inc., San Francisco (1990)
20. Cun, L., Boser, B., Denker, J.S., Henderson, D., Howard, R.E., Hubbard, W., Jackel, L.D.: Handwritten digit recognition with a back-propagation network. In: Advances in Neural Information Processing Systems, pp. 396–404. Morgan Kaufmann, San Mateo (1990)
21. Leung, M.T., Engeler, W.E., Frank, P.: Fingerprint processing using back propagation neural networks. Proc. Int. Joint Conf. Neural Netw. **1**, 15–20 (1990)

22. Gorman, R.P., Sejnowski, T.J.: Analysis of hidden units in a layered network trained to classify sonar targets. Neural Netw. **1**(1), 75–89 (1988)
23. Welch, R.M., Sengupta, S.K., Goroch, A.K., Rabindra, P., Rangaraj, N., Navar, M.S.: Polar cloud and surface classification using AVHRR imagery: An intercomparison of methods. J. Appl. Meteorol. **31**(5), 405–420 (1992)
24. Pomerleau, D.A.: Neural Network Perception for Mobile Robot Guidance. Kluwer, Boston (1993)
25. Poli, R., Cagnoni, S., Livi, R., Coppini, G., Valli, G.: A neural network expert system for diagnosing and treating hypertension. Computer **24**(3), 64–71 (1991)
26. Fujita, H., Katafuchi, T., Uehara, T., Nishimura, T.: Application of artificial neural network to computer-aided diagnosis of coronary artery disease in myocardial SPECT bull's-eye images. J. Nucl. Med. **33**(2), 272–276 (1992)
27. Shea, P.M., Liu, F.: Operational experience with a neural network in the detection of explosives in checked airline luggage. Proc. Int. Joint Conf. Neural Netw. **2**, 175–178 (1990)
28. Tam, K.Y., Kiang, M.Y.: Managerial applications of neural networks: the case of bank failure predictions. Manage. Sci. **38**(7), 926–947 (1992)
29. Hutchinson, J.M.: A radial basis function approach to financial time series analysis. Ph.D. thesis, Department of Electrical Engineering and Computer Science, Massachusetts Institute of Technology, Berkeley (1994)
30. Hao, G., Lai, K.K., Tan, M.: A neural network application in personnel scheduling. Ann. Oper. Res. **128**, 65–90 (2004)
31. Wang, W.: An inventory optimization model based on BP neutral network. In: Proceedings of IEEE 2nd International Conference on Software Engineering and Service Science (ICSESS), pp. 415–418 (2011)
32. Haykin, S.: Kalman Filtering and Neural Networks. Wiley., New York (2002)
33. Ahn, C.W., Ramakrishna, R.S., Kang, C.G., Choi, I.C.: Shortest path routing algorithm using hopfield neural network. Electron. Lett. **37**(19), 1176–1178 (2001)
34. Hippert, H.S., Pedreira, C.E., Souza, R.C.: Neural networks for short-term load forecasting: a review and evaluation. IEEE Trans. Power Syst. **16**(1), 44–55 (2001)
35. Dorffner, G.: Neural networks for time series processing. Neural Netw. World **6**, 447–468 (1996)
36. Stuart, I., Cha, S.-H., Tappert, C.: A neural network classifier for junk e-mail. In: Marinai, S. Dengel, A. (eds.) Document Analysis Systems—VI. Lecture Notes in Computer Science, vol. 3163, pp. 442–450. Springer,Heidelberg (2004)
37. Mohamed, S., Rubino, G.: A study of real-time packet video quality using random neural networks. IEEE Trans. Circuits Syst. Video Technol. **12**(12), 1071–1083 (2002)
38. Misra, J., Saha, I.: Artificial neural networks in hardware: a survey of two decades of progress. Neurocomputing **74**(1–3), 239–255 (2010)
39. Denby, B.: Neural networks in high energy physics: a ten year perspective. Comput. Phys. Commun. **119**(2–3), 219–231 (1999)
40. Weeks, M., Freeman, M., Moulds, A., Austin, J.: Developing hardware-based applications using PRESENCE-2. In: Proceedings of Perspectives in Pervasive Computing, pp. 107–114 (2005)
41. Austin, J.: The cellular neural network associative processor, C-NNAP. In: Proceedings of Fifth International Conference on Image Processing and its Applications, pp. 622–626 (1995)
42. Harrison, R.R.: A biologically inspired analog IC for visual collision detection. IEEE Trans. Circuits Syst. I Regul. Pap. **52**(11), 2308–2318 (2005)
43. Liu, J., Brooke, M.: A fully parallel learning neural network chip for real-time control. In: Proceedings of International Joint Conference on Neural Networks (IJCNN'99), Vol. 4, pp. 2323–2328 (1999)
44. Reyneri, L.M., Chiaberge, M., Zocca, L. CINTIA: a neuro-fuzzy real time controller for low power embedded systems. In: Proceedings of the Fourth International Conference on Microelectronics for Neural Networks and Fuzzy Systems, pp. 392–403 (1994)

45. Camalie: Box 2—an analog audio synthesizer. http://www.camalie.com/MusicBox2/guide.doc (2012). Accessed 30 June 2012
46. Kim, D., Kim, H., Kim, H., Han, G., Chung, D.: A SIMD neural network processor for image processing. In Wang, J., Liao, X.-F., Yi, Z. (eds.) Advances in Neural Networks Lecture Notes in Computer Science, vol. 3497, pp. 815–815. Springer, Heidelberg (2005)
47. Small, J.S.: General-purpose electronic analog computing: 1945–1965. IEEE Ann. Hist. Comput. 15(2), 8–18 (1993)
48. Cowan, G.E.R., Melville, R.C., Tsividis, Y.P.: A VLSI analog computer/digital computer accelerator. IEEE J. Solid State Circ. 41(1), 42–53 (2006)
49. Holler, M., Tam, S., Castro, H., Benson, R.: An electrically trainable artificial neural network (ETANN) with 10240 'floating gate' synapses. Proc. Int. Joint Conf. Neural Netw. 2, 191–196 (1989)
50. Dias, F.M., Antunes, A., Mota, A.M.: Artificial neural networks: a review of commercial hardware. Eng. Appl. Artif. Intell. 17(8), 945–952 (2004)
51. Mumford, M.L., Andes, D.K., Kern, L.L.: The Mod 2 neurocomputer system design. IEEE Trans. Neural Netw. 3(3), 423–433 (1992)
52. Liu, J., Brooke, M.A., Hirotsu, K.: A CMOS feedforward neural-network chip with on-chip parallel learning for oscillation cancellation. IEEE Trans. Neural Netw. 13(5), 1178–1186 (2002)
53. Milev, M., Hristov, M.: Analog implementation of ANN with inherent quadratic nonlinearity of the synapses. IEEE Trans. Neural Netw. 14(5), 1187–1200 (2003)
54. Brown, B., Yu, X., Garverick, S.: Mixed-mode analog VLSI continuous-time recurrent neural network. In: Proceedings of International Conference on Circuits, Signals and Systems, pp. 104–108 (2004)
55. Masa, P., Hoen, K., Wallinga, H.: A high-speed analog neural processor. IEEE Micro 14(3), 40–50 (1994)
56. Lehmann, T., Bruun, E., Dietrich, C.: Mixed analog/digital matrix-vector multiplier for neural network synapses. Analog Integr. Circ. Sig. Process. 9, 55–63 (1996)
57. Kosko, B.: Bidirectional associative memories. IEEE Trans. Syst. Man Cybern. B Cybern. 18(1), 49–60 (1988)
58. Ackley, D.H., Hinton, G.E., Sejnowski, T.J.: A learning algorithm for boltzmann machines. Cogn. Sci. 9(1), 147–169 (1985)
59. Kohring, G.A.: On the Q-state neuron problem in attractor neural networks. Neural Netw. 6(4), 573–581 (1993)
60. Xu, Z.-B., Hu, G.-Q., Kwong, C.-P.: Asymmetric hopfield-type networks: theory and applications. Neural Netw. 9(3), 483–501 (1996)
61. Bhardwaj, S., Jayadeva.: Sequential chaotic annealing neural network for cdma multiuser detection. In: Proceedings of the 9th International Conference on Neural Information Processing (ICONIP'02), vol. 5, pp. 2176–2180 (2002)
62. Wilson, G.V., Pawley, G.S.: On the stability of the travelling salesman problem algorithm of hopfield and tank. Biol. Cybern. 58(1), 63–70 (1988)
63. Rahman, S.A.: A nonlinear synapse neural network and its applications. Ph.D. thesis, Department of Electrical Engineering, Indian Institute of Technology, Delhi (2007)
64. Rahman, S.A., Jayadeva, Dutta Roy, S.C.: Neural network approach to graph colouring. Electron. Lett. 35(14), 1173–1175 (1999)
65. Jayadeva, Rahman, S.A.: A neural network with O(N) neurons for ranking N numbers in O(1/N) time. IEEE Trans. Circuits Syst. I Regul. Pap. 51(10), 2044–2051 (2004)
66. Collins, D.R., Penz, P.A.: Considerations for neural network hardware implementations. In: Proceedings of IEEE International Symposium on Circuits and Systems, Portland, pp. 834–836 (1989)
67. Smith, K.A.: Neural networks for combinatorial optimization: a review of more than a decade of research. INFORMS J. Comput. 11(1), 15–34 (1999)
68. Pershin, Y.V., Di Ventra, M.: Experimental demonstration of associative memory with memristive neural networks. Neural Netw. 23(7), 881–886 (2010)

69. Dinu, A., Cirstea, M.N., Cirstea, S.E.: Direct neural-network hardware-implementation algorithm. IEEE Trans. Industr. Electron. **57**(5), 1845–1848 (2010)
70. Lorrentz, P., Howells, G., McDonald-Maier, K.: An FPGA based adaptive weightless neural network hardware. In: Proceedings of NASA/ESA Conference on Adaptive Hardware and Systems (AHS), Noordwijk, pp. 220–227 (2008)

Chapter 2
Background

To facilitate a better understanding of the material embodied in the subsequent chapters, an overview of the necessary background information is presented in this chapter. It includes a discussion on the Hopfield Neural Network (HNN) including comments on feasibility and limitations of neural network solutions based on the Hopfield approach. Thereafter, another recurrent architecture—the **NO**n-linear **Sy**napse **N**eural **N**etwork (NOSYNN), is explained. Applications in the domain of combinatorial optimization for which the NOSYNN has already been applied, *viz.* sorting of numbers and graph colouring, are also discussed and it is shown that the performance of the NOSYNN based graph colouring network can be improved by a simple modification in the existing network [1]. The chosen problems on which the new (voltage- and mixed-mode) hardware implementations of the NOSYNN have been tested are then explained. Thereafter, an overview of available neural network based solutions, for the problems considered in the book, is presented.

2.1 Hopfield Neural Network

In his seminal paper of 1982, John Hopfield described a new way of modelling a system of neurons capable of performing 'computational' tasks [2]. This approach was by virtue of the fact that a large class of logical problems arising from real world situations could readily be formulated as optimization problems, which in turn essentially means the search for the best solution. Hopfield took his cue from the immense amount of computational powers exhibited by the nervous system of human beings since a massive amount of sensory data is continuously being processed. This coupled with the fact that the human brain churns out answers to recognition problems in relatively small time frames, of the order of milliseconds, led Hopfield to believe (correctly) that most digital computers would fail to provide this combination of computational power and speed. Research is neuroscience has still not fully explained the reason as to how exactly the biophysical properties of neurons and the organization

M. S. Ansari, *Non-Linear Feedback Neural Networks*,
Studies in Computational Intelligence 508, DOI: 10.1007/978-81-322-1563-9_2,
© Springer India 2014

of neurons in the brain combine to give the brain its unmatched computing power and speed. What is known, however, is that the answer may lie in parallel processing. The large degree of interconnectivity between the neurons, which are essentially simple processors, is believed to be the driving force behind the huge computational power of the overall system. Another important feature is that the biological neural system works in an analog mode, with each neuron receiving and summing the outputs of hundreds (sometimes thousands) of neurons in order to compute its graded output, which is then duly fed to some other neuron. With the analog nature comes a disadvantage: although much faster, analog summation is less accurate than its digital counterpart. This limitation has little effect on the performance of the overall system since the tasks which the biological neural system is expected to solve seldom require an exact solution. a 'good' solution (which may not be 'exact') is generally what is required and accepted. Hopfield realized that the biological neural system is able to provide a rapid solution to computationally intensive problems by utilizing the collective analog computation circuits because all the neurons continuously and simultaneously change their states in a parallel fashion. He also correctly associated a non-linear behaviour with the neuron and proceeded to model them using electronic amplifiers. A detailed explanation of Hopfield's neural modelling shall appear later in this chapter.

The Hopfield Neural Network (HNN) initially emerged as a means of exhibiting a content-addressable memory (CAM). A standard CAM must be capable of retrieving a complete item from the system's memory when presented with only sufficient partial information. Hopfield showed that his model was not only capable of correctly yielding an entire memory from any portion of sufficient size, but also included some capacity for generalization, familiarity recognition, categorization, error correction, and time-sequence retention. The Hopfield network, as described in [2–5], comprises of a fully interconnected system of n computational elements or neurons. The strength of the connection, or weight, between neuron i and neuron j is determined by W_{ij}, which may be positive or negative depending on whether the neurons act in an excitatory or inhibitory manner. The internal state of each neuron u_i is equivalent to the weighted sum of the external states of all connecting neurons. The external state of neuron i is given by v_i, with $-1 \leq v_i \leq 1$. An external current input, i_i, to each neuron is also incorporated.

The relationship between the internal state of a neuron and its output level in this continuous Hopfield network is determined by an activation function $g_i(u_i)$. Commonly, this activation function is given by

$$v_i = g_i(u_i) = \tanh\left(\frac{u_i}{T}\right) \tag{2.1}$$

where T is a parameter used to control the gain (or slope) of the activation function. A typical plot of the activation function g_i is presented in Fig. 2.1. In the biological neural system, on which the Hopfield network is based, u_i lags behind the instantaneous outputs, v_j, of the other neurons because of the input capacitance, C_i, of the cell membrane, the trans-membrane resistance R_i, and the finite impedance

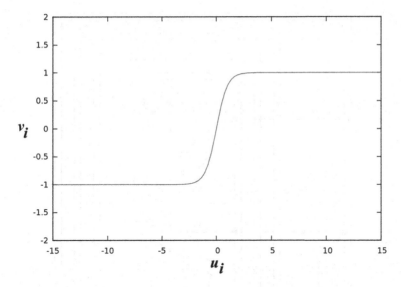

Fig. 2.1 Typical plot of the activation function of a neuron in Hopfield network

$R_{ij}(= W_{ij}^{-1})$ between the output v_j and the cell body of neuron i. Thus, the following resistance-capacitance differential equation determines the rate of change of u_i, and hence the time evolution of the continuous Hopfield network:

$$C_i \frac{du_i}{dt} = \sum_j W_{ij} v_j - \frac{u_i}{R_i} + i_i \tag{2.2}$$

$$u_i = g_i^{-1}(v_i) \tag{2.3}$$

The set of Eqs. (2.2) and (2.3) can also be represented by a resistively connected network of electronic amplifiers shown in Fig. 2.2 with the ith neuron of the network of Fig. 2.2 shown in Fig. 2.3. The synapse (or weight) between two neurons is now defined by a conductance W_{ij}, which connects one of the two outputs of amplifier j to the input of amplifier i. The connection is made with a resistor of value $R_{ij}(= W_{ij}^{-1})$. Figure 2.3 also includes an input resistance, R_{pi}, for amplifier i. This R_{pi} is present implicitly in Eq. (2.2) as being included in R_i which is a parallel combination of R_{pi} and R_{ij}:

$$\frac{1}{R_i} = \frac{1}{R_{pi}} + \sum_j \frac{1}{R_{ij}} \tag{2.4}$$

Node equation for node 'A' gives the equation of motion of the ith neuron of Fig. 2.3 as

$$C_{pi} \frac{du_i}{dt} = \frac{v_1}{R_{i1}} + \frac{v_2}{R_{i2}} + \cdots + \frac{v_n}{R_{in}} - u_i \left[\frac{1}{R_i} \right] + i_i \tag{2.5}$$

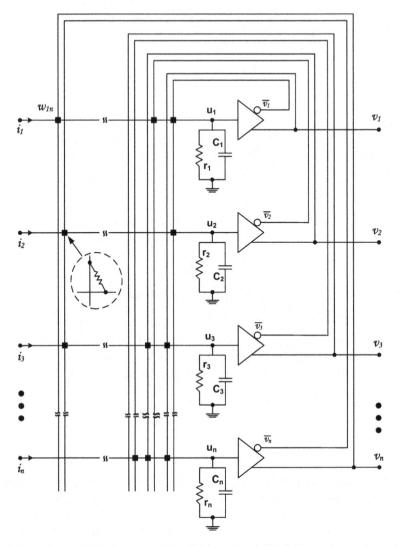

Fig. 2.2 An electronic circuit representation of the continuous Hopfield neural network

which can be written as

$$C_{pi}\frac{du_i}{dt} = \sum_j \left[\frac{v_j}{R_{ij}}\right] - u_i\left[\frac{1}{R_i}\right] + i_i \qquad (2.6)$$

It can be seen that (2.6) is the same as (2.2) with the weights being implemented by resistances. For simplicity, each neuron/amplifier is assumed to be identical, so that

Fig. 2.3 ith neuron of
Hopfield neural network

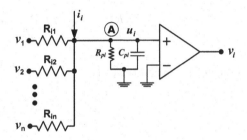

$$R_i = R; \; C_i = C; \; g_i = g; \tag{2.7}$$

Dividing (2.2) by C and redefining W_{ij}/C and i_i/C to be W_{ij} and i_i, respectively, we arrive at the normalized equations of motion:

$$\frac{du_i}{dt} = \sum_j W_{ij} v_j - \frac{u_i}{\tau} + i_i \tag{2.8}$$

$$\tau = RC \tag{2.9}$$

τ is the value of the time constant of the amplifiers, and without loss of generality can be assigned a value of unity, provided the time step of the discrete time simulation of (2.8) is considerably smaller than unity. Although this 'neural' computational network has been described in terms of an electronic circuit, it has been shown [5] that biological models with action potentials and excitatory and inhibitory synapses can compute in a similar fashion to this electrical hardware.

To ascertain the stability of the system described by Eqs. (2.2) and (2.3), Hopfield employed the concept of a so called 'generalized computational energy function,' E, which is essentially a Lyapunov function associated with the network which guarantees convergence to stable states. For the continuous Hopfield network, the ith neuron of which is shown in Fig. 2.3, the gradient of the energy function is related to the evolution of the neuronal state as

$$\frac{\partial E}{\partial V_i} = -C_{pi} \frac{du_i}{dt}; \quad \text{for all } i \tag{2.10}$$

Using (2.6) and (2.10), the energy function corresponding to the network of Fig. 3.4 can be written as

$$E = -\frac{1}{2} \sum_i \sum_j W_{ij} v_i v_j - \sum_i i_i v_i + \sum_i \frac{1}{R_i} \int_0^{v_i} g_i^{-1}(\mathbf{v}) \, d\mathbf{v} \tag{2.11}$$

Provided the matrix of weights $\mathbf{W}$ is symmetric (although Vidyasagar [6] has shown that convergence is still possible under some asymmetric conditions), the time derivative of E is:

$$\frac{dE}{dt} = \sum_{i=1}^{N} \frac{\partial E}{\partial v_i} \frac{dv_i}{dt} = \sum_{i=1}^{N} \frac{\partial E}{\partial v_i} \frac{dv_i}{du_i} \frac{du_i}{dt} \tag{2.12}$$

Using (2.10) in (2.12) we get

$$\frac{dE}{dt} = -\sum_{i=1}^{N} C_i \left(\frac{du_i}{dt}\right)^2 \frac{dv_i}{du_i} \tag{2.13}$$

The transfer characteristics of the opamp used in Fig. 2.3 implements the activation function of the neuron. With u_i being the internal state at the non-inverting terminal, it is monotonically increasing as shown in Fig. 2.1, and therefore,

$$\frac{dv_i}{du_i} \geq 0 \tag{2.14}$$

thereby resulting in

$$\frac{dE}{dt} \leq 0 \tag{2.15}$$

with the equality being valid for

$$\frac{du_i}{dt} = 0 \tag{2.16}$$

Together with the boundedness of E, (2.15) shows that under control of the differential equation (2.6), E decreases and converges to a minimum, at which it stays.

Hopfield, along with Tank, realized that the analog nature of the neurons and the parallel processing of the updating procedure could be combined to create a rapid and powerful solution technique. They presented several applications of the Hopfield neural network including Content Addressable Memory (CAM), solution of Travelling Salesman Problem (TSP), analog-to-digital conversion, signal decision and solution of Linear Programming Problem (LPP). The underlying idea in most of these applications was the fact that solution to specific optimization problems could be obtained by selecting weights and external inputs that appropriately represent the function to be minimized along with the desired states of the problem. In other words, the 'energy function' corresponding to the network is made equivalent to the objective function of the optimization problem that needs to be minimized, while the constraints of the problem are included in the energy function as penalty terms. Once a suitable energy function has been chosen, the network parameters (weights and inputs) can be inferred by comparison with the standard energy function given by (2.11). The weights of the continuous Hopfield network, W_{ij}, are the coefficients of the quadratic terms $v_i v_j$, and the external inputs, i_i, are the coefficients of the linear terms v_i in the chosen energy function. The network can then be initialized by setting the activity level v_i of each neuron to a small random perturbation. From its initialized state, asynchronous updating of the network will then allow a minimum energy state to be attained.

However, these stable states may not necessarily correspond to *feasible* or *good* solutions of the optimization problem, and this is one of the major pitfalls of the Hopfield neural network. Because the energy function comprises several terms (each of which is competing to be minimized), there are many local minima, and a tradeoff exists between which terms will be minimized. An infeasible solution to the problem will arise when at least one of the constraint penalty terms is non-zero. If this occurs, the objective function term is generally quite small, because it has been minimized to the detriment of the constraint terms, thus the solution is "good" but not feasible. Alternatively, all constraints may be satisfied, but a local minimum may be encountered that does not globally minimize the objective function, in which case the solution is feasible but not "good." To overcome such a situation, a penalty parameter can be increased to force its associated term to be minimized, but this generally has the unwanted effect of causing other terms to be increased. The solution to this trade-off problem is to find the optimal values of the penalty parameters that balance the terms of the energy function and ensure that each term is minimized. Only then will the constraint terms be zero (a feasible solution), and the objective function be also minimized (a "good" solution).

Further, due to certain omissions in Hopfield and Tank's paper [3] relating to termination criteria, simulation procedure for differential equations, etc. it has been quite difficult to reproduce the results for TSP. Wilson and Pawley were the first to report this discrepancy and suggested many modifications in the original Hopfield neural network [7]. Over time, two distinct approaches to improve the original Hopfield neural network have been developed. The first approach consists of rewriting the energy function to eliminate the trade-offs between valid and good solutions. The second technique is to accept the Hopfield energy function as such while searching for ways to optimally select the penalty parameters. A systematic method for selection of parameters based on the analysis of dynamic stability of valid solutions has been developed by Kamgar-Parsi and Kamgar-Parsi [8].

Efforts to obtain better results by modification of energy function have proved to be more fruitful. Significant contributions in this regard include the *valid subspace approach* of Aiyer et al. [9] and the subsequent work by Gee [10]. Other approaches of modifying the energy function, specific to the TSP problem, have also been reported [11, 12]. Deterministic methods like the "divide and conquer" technique of Foo [13], "rock and roll" perturbation method of Lo [14], and the use of alternative neuron models within the Hopfield network such as the winner-take-all neurons [15], to improve the quality of solution have also been presented.

Although the fundamental worth of Hopfield network is beyond contention, the never ending quest to improve solution quality has resulted in new models of neuron dynamics being investigated. Chaotic neural networks [16] and hybridization of neural networks with meta-heuristics such as simulated annealing and genetic algorithms [17] have been offered as alternatives to the Hopfield neural network.

Due to the limitations mentioned above as well as the requirement of a large number of neurons and interconnection weights, attempts to apply Hopfield networks in obtaining working hardware for practical applications have not proved to be encouraging. For instance, to solve a N-city TSP problem, a Hopfield network requires N^2 neurons and $O(N^4)$ weights. This translates to the requirement of 100 neurons and 10000 weights for a 10-city TSP problem thereby clearly indicating that the Hopfield network is not amenable to a hardware realization. Alternative neural network architectures were therefore explored, some of which are mentioned above. Another neural architecture which employs non-linear feedback, as opposed to linear feedback in the Hopfield network, has been proposed as an alternative to the Hopfield network [18]. It has been shown that the architecture facilitates more efficient hardware implementations for combinatorial problems as compared to the Hopfield network [18]. The details of the non-linear feedback neural network are discussed next.

2.2 Nonlinear Synapse Neural Network

A recurrent neural network architecture, termed as Nonlinear Synapse Neural Network (NOSYNN), has been proposed as an alternative to the Hopfield network [18]. In a manner similar to the Hopfield network, the NOSYNN is comprised of a single layer of neurons and a fully interconnected feedback structure. The most important difference between the two is in the nature of feedback. While the Hopfield network employs linear synapses, the NOSYNN uses nonlinear synapses to which input signals are fed back from the neuron outputs. The nonlinearity in the synapse is implemented by using comparators in the circuit realization of the network. The provision of a direct feedback from a neuron to itself is also present as opposed to the original Hopfield network in which self-interactions in the neurons were not allowed.

Figure 2.4 presents the schematic of the NOSYNN and the corresponding electronic circuit implementation of NOSYNN is presented in Fig. 2.5. The internal state and output of the ith neuron are denoted by u_i and V_i respectively. S_{ij} represent the synapses giving feedback from the jth neuron to the ith neuron. The output of S_{ij} is denoted by x_{ij}, which is a non-linear function of V_i and V_j, and is given by

$$x_{ij} = V_m \tanh \beta(V_j - V_i) \tag{2.17}$$

where $\pm V_m$ are the maximum output levels of the comparator and β is the gain of the comparator. For a high gain comparator, the transfer characteristics are practically indistinguishable from a signum function, as shown in Fig. 2.6. Individual resistors are used to assign different weights to the comparator outputs. To introduce a weight W_{ij} in the feedback connection from the jth neuron to the ith neuron, a resistance R_{ij} is employed where

$$R_{ij} = \frac{1}{W_{ij}} \tag{2.18}$$

Another noteworthy feature of the NOSYNN is the possibility of obtaining negative weights by simply interchanging the comparator inputs. This is possible due

Fig. 2.4 Schematic of the
NOSYNN architecture

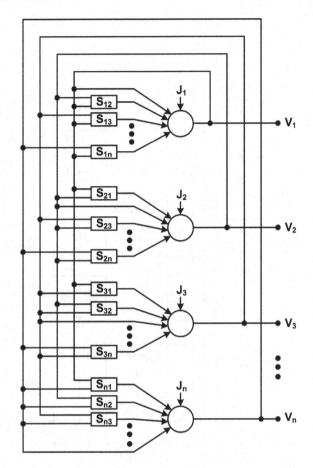

to fact that the comparator transfer characteristics are given by the tanh(.) function which is an odd function. This eliminates the need of generating inverted amplifier outputs for implementing negative weights, as is required in the Hopfield network.

The equation of motion of the ith neuron may be obtained by applying KCL at the node u_i, which is the internal state of the ith neuron. From Fig. 2.5, the following differential equation is obtained.

$$C_{pi}\frac{du_i}{dt} + \frac{u_i}{r_{pi}} = I_i + \frac{V_i - u_i}{R_{ii}} + \sum_{\substack{j=1 \\ j \neq i}}^{n} \frac{x_{ij} - u_i}{R_{ij}} \tag{2.19}$$

Using (2.19) and (2.17), we have

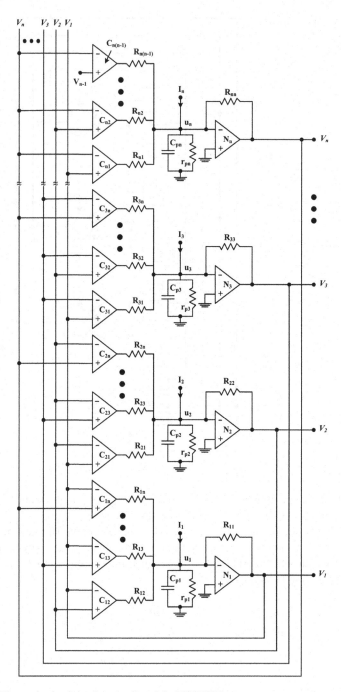

Fig. 2.5 Electronic circuit implementation of the NOSYNN

Fig. 2.6 Comparator outputs in the high gain limit $(\beta \to \infty)$

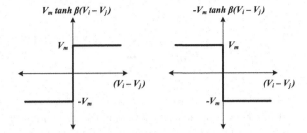

$$C_{pi}\frac{du_i}{dt} + \frac{u_i}{r_{pi}} = I_i + \frac{V_i - u_i}{R_{ii}} + \sum_{\substack{j=1 \\ j \neq i}}^{n} \frac{V_m\tanh\beta(V_j - V_i) - u_i}{R_{ij}} \qquad (2.20)$$

which can be further simplified to

$$C_{pi}\frac{du_i}{dt} = I_i + \frac{V_i}{R_{ii}} - \frac{u_i}{R_i} + \sum_{\substack{j=1 \\ j \neq i}}^{n} \frac{V_m\tanh\beta(V_j - V_i)}{R_{ij}} \qquad (2.21)$$

where

$$\frac{1}{R_i} = \frac{1}{r_{pi}} + \frac{1}{R_{ii}} + \sum_{\substack{j=1 \\ j \neq i}}^{n} \frac{1}{R_{ij}} \qquad (2.22)$$

Therefore, R_i is equivalent to the parallel combination of all the resistances connected at the input of the neuron i.e. to the node at internal state u_i. At the steady state, the internal and output states of the neuron may be obtained by putting the left-hand side of (2.21) equal to 0, and are given as

$$u_i = I_i + \frac{V_i}{R_{ii}} + \sum_{\substack{j=1 \\ j \neq i}}^{n} \frac{V_m\tanh\beta(V_j - V_i)}{R_{ij}} \qquad (2.23)$$

$$V_i = -\tanh(\lambda u_i) \qquad (2.24)$$

where λ is the gain of the operational amplifier used to implement the activation function of the neuron. It is to be noted that both R_i and C_i have been normalized to unity without any loss of generality.

The NOSYNN may be associated with the following Lyapunov function [18]:

$$E = \sum_{i=1}^{n} \left[I_i V_i + \frac{V_i^2}{2R_{ii}} - \frac{1}{2} \sum_{\substack{j=1 \\ j \neq i}}^{n} \frac{V_m \ln \left[\cosh \beta \left(V_j - V_i \right) \right]}{\beta} - \frac{1}{R_i} \int_0^{V_i} u_i \, dV_i \right]$$

(2.25)

It has been assumed that the weights are symmetric i.e. $W_{ij} = W_{ji}, (i, j = 1, 2, \ldots, n)$. Due to the inclusion of nonlinearity in the feedback path (implemented by using comparators), the Lyapunov function for the NOSYNN, as given in (2.25) is considerably different from the corresponding function for the standard Hopfield network, given in (2.11). While the Hopfield network is associated with a Lyapunov function containing quadratic terms, the NOSYNN is characterized by transcendental terms in the Lyapunov function.

The fact that E, given in (2.25) is indeed a valid Lyapunov function for the NOSYNN can be proved as follows. The ith component of the gradient of E is given by

$$\frac{\partial E}{\partial V_i} = I_i + \frac{V_i}{R_{ii}} - \frac{u_i}{R_i} + \sum_{\substack{j=1 \\ j \neq i}}^{n} \frac{V_m \tanh \beta \left(V_j - V_i \right)}{R_{ij}}$$

(2.26)

Comparing (2.21) and (2.26), with C_{pi} normalized to unity, we get

$$\frac{\partial E}{\partial V_i} = \frac{du_i}{dt}$$

(2.27)

The total derivative of E with respect to time is given by

$$\frac{dE}{dt} = \sum_i \frac{\partial E}{\partial V_i} \frac{dV_i}{du_i} \frac{du_i}{dt}$$

(2.28)

From (2.27) and (2.28), we have

$$\frac{dE}{dt} = \sum_i \frac{dV_i}{du_i} \left[\frac{du_i}{dt} \right]^2$$

(2.29)

From (2.24), we have

$$\frac{dV_i}{du_i} \leq 0$$

(2.30)

Using (2.30) in (2.29), we get

$$\frac{dE}{dt} \leq 0$$

(2.31)

which proves that E is non-increasing with time, thereby fulfilling one (of two) condition for a valid Lyapunov function. The second requirement of E being bounded from below can be justified by considering the fact that since the amplifier outputs saturate, the neuron states are bounded from below as well as from above. Therefore, E, which is a function of the neuron outputs and is bounded for bounded arguments, will also be bounded.

The Lyapunov function associated with feedback neural networks, like the Hopfield network and the NOSYNN, has been extensively used to solve hard optimization problems. Such Lyapunov function based networks are guaranteed to settle at a stable output corresponding to one of its minima. If a particular network is so designed that the minima of its associated Lyapunov function (also referred to as the 'energy' function) correspond to solution states of a given optimization problem, then the network can be viewed as a dynamical computing system for solving the specific problem. The applicability of NOSYNN in solving combinatorial optimization problems like graph colouring and ranking, by proper selection of the 'energy' function, has been demonstrated [1, 19]. It is to be mentioned however that although the network of [1] yielded better colouring results alongwith reduced circuit complexity, the performance of the network can be further improved by incorporating a slight alteration in the circuit. The details of such a modified NOSYNN based graph coloring network are presented later in this chapter.

2.2.1 Sorting of Numbers Using NOSYNN

Ordering of a set of numbers based on their relative magnitudes, which is analogous to sorting, is a fundamental operation in computing. Almost all computing tasks including, but not limited to, VLSI design, digital signal processing, network communications, database management and data processing need to perform the fundamental operation of sorting. A multitude of sorting techniques has been reported in the technical literature. Considerable work has been carried out towards the development of parallel sorting methods that minimize both the sorting time and the number of processors. In recent years, neural networks, by virtue of their inherent parallelism, have also been applied to sorting of numbers. In this section, a NOSYNN based neural circuit capable of ranking numbers according to their respective magnitudes is discussed.

The ith neuron of the NOSYNN based circuit for ranking of numbers is shown in Fig. 2.7 from where it is evident that the output of the ith neuron is denoted by V_i [19]. In the complete neural network, each neuron receives inputs from all others, and its output is connected to the inputs of all other neurons, thereby creating a fully interconnected feedback structure. The parasitic capacitance C_{pi} and the parasitic resistance r_{pi} are included to model the dynamic nature of the operational amplifier. Since the nature of feedback is non-linear, the synapses are implemented using opamp based voltage-mode comparators. For sorting n numbers, the values of the resistances are obtained as follows. First an appropriate value of resistance is selected for the

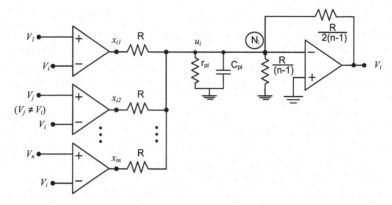

Fig. 2.7 ith neuron of the NOSYNN based neural network for ranking of numbers

resistors marked 'R'. Thereafter, the resistances labelled as $R/(n-1)$ and $R/2(n-1)$ may be found for the given n. Furthermore, since for moderate to large values of n the value of the resistance $R/(n-1)$ would turn out to be small, the effect of the parasitic resistance r_{pi} may be neglected.

The working of the circuit of Fig. 2.7 can be understood by first understanding the 'energy function' associated with the circuit for ranking two numbers. In that case, two neurons would be required and only two comparators would be needed. The inputs to the first comparator would be V_2 and V_1 at the non-inverting and the inverting terminals respectively. The inputs to the second comparator would be V_1 and V_2 at the non-inverting and the inverting terminals respectively. The values of the resistances can be calculated as

$$\frac{R}{n-1} = R \ (since \ n = 2) \tag{2.32}$$

and

$$\frac{R}{2(n-1)} = \frac{R}{2} \ (since \ n = 2) \tag{2.33}$$

The output of the first comparator, x_1 would be given by

$$x_1 = V_m \tanh \beta(V_2 - V_1) \tag{2.34}$$

where β is the open-loop gain of the comparator and is typically large. Similarly, the output of the second comparator can be written as

$$x_2 = V_m \tanh \beta(V_1 - V_2) \tag{2.35}$$

Application of KCL at the input of the first neuronal amplifier (corresponding to the node labelled as N_i in the generalized diagram in Fig. 2.7) results in

$$\frac{x_1 - u_1}{R} + \frac{V_1 - u_1}{R/2} = C_{p1}\frac{du_1}{dt} + \frac{u_1}{r_{p1}} + \frac{u_1}{R} \tag{2.36}$$

Rearranging, we get

$$C_{p1}\frac{du_1}{dt} = \frac{x_1}{R} + \frac{V_1}{R/2} - u_1\left[\frac{1}{r_{p1}} + \frac{2}{R} + \frac{1}{R} + \frac{1}{R}\right] \tag{2.37}$$

Using (2.34) in (2.36) results in

$$C_{p1}\frac{du_1}{dt} = \frac{V_m \tanh \beta(V_2 - V_1)}{R} + \frac{V_1}{R/2} - u_1\left[\frac{1}{r_{p1}} + \frac{2}{R} + \frac{1}{R} + \frac{1}{R}\right] \tag{2.38}$$

which can be simplified to

$$C_{p1}\frac{du_1}{dt} = \frac{V_m \tanh \beta(V_2 - V_1)}{R} + \frac{V_1}{R/2} - \frac{u_1}{r_{eff1}} \tag{2.39}$$

where

$$\frac{1}{r_{eff1}} = \frac{1}{r_{p1}} + \frac{2}{R} + \frac{1}{R} + \frac{1}{R} \tag{2.40}$$

A similar analysis when performed for the second neuron would yield

$$C_{p2}\frac{du_2}{dt} = \frac{V_m \tanh \beta(V_1 - V_2)}{R} + \frac{V_2}{R/2} - u_2\left[\frac{1}{r_{p2}} + \frac{2}{R} + \frac{1}{R} + \frac{1}{R}\right] \tag{2.41}$$

which can be simplified to

$$C_{p2}\frac{du_2}{dt} = \frac{V_m \tanh \beta(V_1 - V_2)}{R} + \frac{V_2}{R/2} - \frac{u_2}{r_{eff2}} \tag{2.42}$$

where

$$\frac{1}{r_{eff2}} = \frac{1}{r_{p2}} + \frac{2}{R} + \frac{1}{R} + \frac{1}{R} \tag{2.43}$$

Also, for such dynamical systems like the one shown in Fig. 2.7, the gradient of the energy function E is related to the time evolution of the network as given by (2.44).

$$\frac{\partial E}{\partial V_i} = C_{pi}\frac{du_i}{dt} \tag{2.44}$$

and since we are analysing a 2 neuron system,

$$\frac{\partial E}{\partial V_1} = C_{p1}\frac{du_1}{dt} \tag{2.45}$$

$$\frac{\partial E}{\partial V_2} = C_{p2}\frac{du_2}{dt} \tag{2.46}$$

From (2.45) and (2.46), the energy function associated with a 2 neuron network for ranking of 2 numbers can be obtained as

$$E = \frac{V_1{}^2}{R/2} + \frac{V_2{}^2}{R/2} - \frac{V_m}{\beta R}\ln\cosh\beta(V_2 - V_1) \tag{2.47}$$

It may be mentioned that the energy function given above consists only of the 'significant' terms and the terms which have only 'insignificant' and negligible effect (terms corresponding to the terms with u_1 and u_2 in (2.39) and (2.42) respectively) are not included. The different terms in the energy function obtained above in (2.47), work in tandem to drive the neuronal output voltages away from each other. For the case of 2 numbers being ranked by a 2 neuron system, the 'away' states are $+V_m/2$ and $-V_m/2$.

For a larger network with n neurons capable of ranking n numbers, the energy function can be derived as follows. Considering the ith neuron, as shown in Fig. 2.7, the comparator outputs can be given as

$$x_{i1} = V_m\tanh\beta(V_1 - V_i) \tag{2.48}$$

$$x_{ij} = V_m\tanh\beta(V_j - V_i) \tag{2.49}$$

$$\vdots$$

$$x_{in} = V_m\tanh\beta(V_n - V_i) \tag{2.50}$$

Application of KCL at the node maked N_i yields

$$\frac{x_{i1} - u_i}{R} + \frac{x_{ij} - u_i}{R} + \cdots + \frac{x_{in} - u_i}{R} + \frac{V_i - u_i}{R/2(n-1)} = C_{pi}\frac{du_i}{dt}\frac{u_i}{r_{pi}} + \frac{u_i}{R/(n-1)} \tag{2.51}$$

Rearranging, we get

$$C_{pi}\frac{du_i}{dt} = \left[\frac{x_{i1}}{R} + \frac{x_{ij}}{R} + \cdots + \frac{x_{in}}{R}\right] + \frac{V_i}{R/2(n-1)}$$
$$- u_i\left[\frac{1}{r_{pi}} + \frac{1}{R/(n-1)} + \frac{1}{R/2(n-1)} + \frac{n-1}{R}\right] \tag{2.52}$$

which can be written in a simplified form as

$$C_{pi}\frac{du_i}{dt} = \left[\frac{x_{i1}}{R} + \frac{x_{ij}}{R} + \cdots + \frac{x_{in}}{R}\right] + \frac{V_i}{R/2(n-1)} - \left[\frac{u_i}{r_{effi}}\right] \tag{2.53}$$

where

$$\frac{1}{r_{effi}} = \left[\frac{1}{r_{pi}} + \frac{1}{R/(n-1)} + \frac{1}{R/2(n-1)} + \frac{n-1}{R}\right] \qquad (2.54)$$

Substituting the values of $x_{i1}, x_{ij}, \ldots, x_{in}$ in (2.53) from (2.48) through (2.50) yields

$$C_{pi}\frac{du_i}{dt} = V_m \left[\frac{\tanh \beta(V_1 - V_i)}{R} + \frac{\tanh \beta(V_j - V_i)}{R} + \cdots + \frac{\tanh \beta(V_n - V_i)}{R}\right]$$
$$+ \frac{V_i}{R/2(n-1)} - \left[\frac{u_i}{r_{effi}}\right] \qquad (2.55)$$

Now, the energy function associated with the ranking circuit comprising of n neurons can be found by considering the fact that for dynamical systems like the one shown in Fig. 2.7, the gradient of the energy function E is related to the time evolution of the network as given by (2.56).

$$\frac{\partial E}{\partial V_i} = C_{pi}\frac{du_i}{dt} \qquad (2.56)$$

Combining (2.55) with (2.56), we get the ith term in the energy function expression as

$$E_i = -\frac{V_m}{\beta R} \ln \cosh \beta(V_1 - V_i) - \frac{V_m}{\beta R} \ln \cosh \beta(V_j - V_i)$$
$$\cdots - \frac{V_m}{\beta R} \ln \cosh \beta(V_n - V_i) + \frac{V_i^2}{R/(n-1)} \qquad (2.57)$$

The complete energy function can therefore be written as

$$E = \sum_{i=1}^{N} \frac{V_i^2}{R/2(n-1)} - \frac{V_m}{2\beta R} \sum_{i=1}^{N} \sum_{\substack{j=1 \\ j \neq i}}^{N} \ln \cosh \left(\beta \left(V_i - V_j\right)\right)$$

Again, it needs to be pointed out that the above energy function has been written after neglecting the terms corresponding to the u_i terms in (2.55), since the effect of such terms is negligible and is usually neglected.

For the energy function of (2.58) to be a valid energy function, the following conditions need to be satisfied.

- E should have a definite lower bound.
- E should be monotonically decreasing with time.

The first criterion is easily satisfied by the circuit of Fig. 2.7 by virtue of the fact that the operational amplifiers used in the network enforce a definite lower bound on their output voltages, which is equal to the negative biasing voltage. The second criterion can be proved by considering that the time derivative of the energy function is given by

$$\frac{dE}{dt} = \sum_{i=1}^{N} \frac{\partial E}{\partial V_i} \frac{dv_i}{dt} = \sum_{i=1}^{N} \frac{\partial E}{\partial V_i} \frac{dV_i}{du_i} \frac{du_i}{dt} \tag{2.58}$$

Using (2.56) in (2.58) we get

$$\frac{dE}{dt} = \sum_{i=1}^{N} C_i \left(\frac{du_i}{dt} \right)^2 \frac{dV_i}{du_i} \tag{2.59}$$

The transfer characteristics of the opamp used in Fig. 2.7 implements the activation function of the neuron. With u_i being the internal state at the inverting terminal, the transfer characteristics are monotonically decreasing, and therefore,

$$\frac{dV_i}{du_i} \leq 0 \tag{2.60}$$

thereby resulting in

$$\frac{dE}{dt} \leq 0 \tag{2.61}$$

with the equality being valid for

$$\frac{du_i}{dt} = 0 \tag{2.62}$$

Some interesting points about the circuit of Fig. 2.7 need to be pointed out at this stage. Firstly, since the opamp realizing the neuronal amplifier is connected in an inverting configuration, virtual ground will try to enforce a voltage value which is very small, at the node marked N_i. This is the same as asserting that u_i is negligible and from this emanates the reason for neglecting the terms corresponding to u_i in the energy function expression in (2.58).

Another point worth consideration is the presence/absence of the resistance labelled as $R/(n-1)$ connected at the node N_i in the circuit of Fig. 2.9. Since one end of the resistance is physically grounded and the other end is approximately at ground (very low voltage, u_i), there would be only a negligible amount of current flowing through the resistance and the entirety of the current(s) coming from the comparators to the node N_i will therefore pass through the $R/2(n-1)$ resistance connected in the feedback path across the opamp. The grounded resistance may therefore be eliminated in a 'compact' hardware implementation of the circuit.

The mechanism behind the actual assignment of voltage ranks performed by the circuit can be understood more clearly by taking some example numbers in the circuit of Fig. 2.9. For instance, if the output nodes of the operational amplifiers (biased at $\pm V_{CC} = \pm 15V$) are initialized at $V_1 = 1V$, $V_2 = 2V$ and $V_3 = 3V$ (corresponding to the numbers to be ranked being 1, 2 and 3 respectively), the outputs of the two comparators connected to the first neuron will be

$$x_{11} = +15V; \quad x_{12} = +15V;$$

Assuming u_i to be negligible, the total current arriving at the node N_1 from the two comparators would be

$$i_{N1} = 2 \times \frac{15}{R}$$

This current will *not* flow into the grounded resistance $R/(n-1)$ due to reasons mentioned above. Instead, this current will flow through the resistance $R/2(n-1)$ thereby producing an output voltage equal to

$$V_1 = -2 \times \frac{15}{R} \times \frac{R}{4} = -\frac{15}{2} = -\frac{V_{CC}}{2}$$

For the second neuron, the comparator outputs for the above mentioned inputs would be

$$x_{21} = -15V; \quad x_{12} = +15V;$$

The total current arriving at the node N_2 from the two comparators would be

$$i_{N2} = 0$$

This will result in an output voltage equal to

$$V_2 = 0$$

Similarly, for the third neuron, the comparator outputs would be

$$x_{31} = -15V; \quad x_{32} = -15V;$$

The total current arriving at the node N_3 from the two comparators would be

$$i_{N3} = -2 \times \frac{15}{R}$$

This current will result in an output voltage equal to

$$V_1 = +2 \times \frac{15}{R} \times \frac{R}{4} = +\frac{15}{2} = +\frac{V_{CC}}{2}$$

The above discussion correctly verifies the final node voltages obtained for a 3 neuron network which is used to rank three numbers *viz.* 1, 2 and 3. It has been shown above that the output of the first neuron goes to $-V_{CC}/2$ (starting from the initialized state of 1 Volt), the output of the second neuron settles at zero (starting from the initial

Table 2.1 PSPICE simulation results for the NOSYNN based neural network for ranking of two numbers

Numbers to be ranked (given as initial condition at the neuron outputs) V1, V2(V)	Final steady state neuronal output voltages obtained in PSPICE simulations V1, V2 (V)
5, −7	7.406, −7.406
0.1, 0.5	−7.406, 7.406
2, −3	7.406, −7.406
−1, −1.1	7.406, −7.406
5, 5.02	−7.406, 7.406
0.020, 0.025	−7.406, 7.406
0.050, −2	7.406, −7.406

condition of 2 Volts) and the voltage corresponding to the third neuron reaches a steady state value of $+V_{CC}/2$ (starting from the initial 3 Volts). It is readily verified that the final neuronal voltages are in the order of the ranks of the numbers to be sorted, with the largest number being placed at the highest rank i.e. at the highest available voltage slot.

Verification of the operation of the NOSYNN based feedback neural circuit for ranking was carried out using computer simulations done in PSPICE software. It was observed that the network yielded correct results in all the test cases. First, a simple circuit for sorting of two numbers was set up in PSPICE and the network was found to correctly assign a voltage equal to $+V_m/2$ to the larger number and $-V_m/2$ to the smaller number. Some of the test results are presented in Table 2.1. As can be seen from the tabulated results, the neural network assigned the higher of the available choices at the output (i.e. $+V_m/2$) to the larger number. The PSPICE simulations were carried out using $\mu A741$ operational amplifiers biased at ± 15 Volts.

A variety of test cases are included in Table 2.1. While the first set (5, −7) comprises of numbers which are moderately large integers, the second (0.1, 0.5) deals with small positive numbers. The fourth set (−1, −1.1) tests the circuit for both negative numbers. The network was also tested for numbers which were very small in magnitude and also very close together. The set (0.020, 0.025), applied to the ranking circuit resulted in steady state neuronal voltages of V(1) = −7.406 and V(2) = +7.406. This serves to verify that the NOSYNN based neural network for ranking is indeed capable of providing correct sorting for the numbers applied as initial conditions to the network. The results of PSPICE simulations for the last test case in Table 2.1 are shown in Fig. 2.8 from where it can be observed that the circuit takes a time of the order of tens of microseconds to correctly sort the two numbers and provide steady state neuronal output voltages.

Having verified the operation of the ranking network for sorting two numbers, the network was employed to rank three numbers. The circuit to sort three numbers, obtained using the generalized neuron shown in Fig. 2.7, is presented in Fig. 2.9 from where it can be seen that feedback connections are present from the output

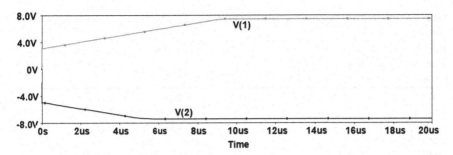

Fig. 2.8 Neuron outputs in the NOSYNN based neural network for ranking of two numbers

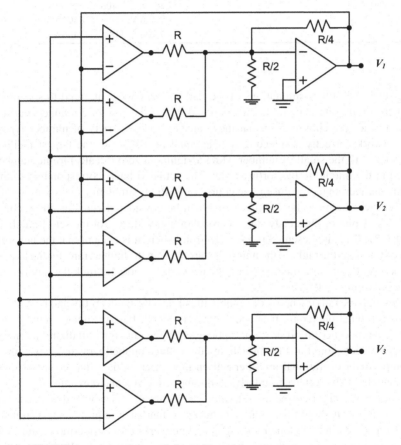

Fig. 2.9 NOSYNN based neural network for ranking of three numbers

of each neuron to the inputs of other neurons, as well to the neuron's own input (self-feedback). As has already been mentioned before, such self interactions are not allowed in the standard Hopfield Neural Network.

Table 2.2 PSPICE simulation result for the NOSYNN based neural network for ranking of three numbers

Numbers to be ranked (given as initial condition at the neuron outputs) V1, V2, V3 (V)	Final steady state neuronal output voltages obtained in PSPICE simulations V1, V2, V3 (V)
2, 0, 1	7.406 V, -7.406 V, 118 μV
$-2, 0, 1$	-7.406 V, 118 μV, 7.406 V
$-5, 3, -1.1$	-7.406 V, 7.406 V, 118 μV
0.20, 0.21, 0.15	118 μV, 7.406 V, -7.406 V
$-0.020, -0.017, -0.025$	118 μ V, 7.406 V, -7.406 V
7, $-7, -5$	7.406 V, -7.406 V, 118 μV
6.53, -5.90, 1.25	7.406 V, -7.406 V, 118 μV
$-3, 0, 4.35$	-7.406 V, 118 μV, 7.406 V
2, 3, 5	-7.406 V, 118 μV, 7.406 V

Results of PSPICE simulations of the ranking network configured to sort three numbers are presented in Table 2.2. As was done for the case of 2 numbers earlier, the test cases are chosen such that they cater to a wide variety of number types. The network correctly assigned a voltage equal to $+V_m/2$ to the largest number and $-V_m/2$ to the smallest number while assigning a zero (ideally) to the number placed in the middle in the sorting order. The actual voltage, corresponding to the middle number, obtained during the course of PSPICE simulations was not exactly zero and was found to be $118\,\mu$V, which can be considered zero for all practical purposes. A plot of the steady state neuronal output voltages for the set of numbers (0.20, 0.21, 0.15) is presented in Fig. 2.10 from where it is evident that the network correctly assigns the ranks to the numbers starting from the highest rank for the largest number. As before, the time taken by the network to correctly rank the numbers is in the proximity of $10\,\mu$s.

Although the ranking network suitably illustrates the applicability and benefits of analog computation, it must be noted that the NOSYNN based ranking circuit would need to be modified in order to incorporate arrangements for initializing the node voltages, which would involve initializing the voltage on the (parasitic) capacitor at the output of the operational amplifier with an appropriate initial value (corresponding to the number to be sorted). Similarly, additional arrangements for reading the steady state voltages, and for correcting for voltage and current offsets would be needed in actual practice. It should be mentioned however, that such requirements also need to be considered in many other sorting neural networks too. Furthermore, since the steady state output node voltages of the neurons are separated equally between the voltages $-V_m/2$ to $+V_m/2$, each output voltage in a set-up for sorting n numbers would differ from its nearest neighbour by (V_m/n). For high values of n, this can cause the output node voltages to become indistinguishable from each other, and in the presence of noise, the ranked orders would be impossible to discern. In other words, an upperbound on n would be set by the noise floor.

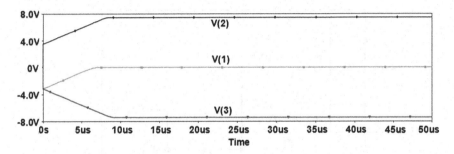

Fig. 2.10 Neuron outputs in the NOSYNN based neural network for ranking of three numbers

2.2.2 *Graph Colouring Using NOSYNN*

In graph theory, the Graph Colouring Problem (GCP) is a special case of the Graph
Labelling Problem (GLP). GCP is essentially an assignment of labels, traditionally
referred to as 'colours,' to elements of a given graph subject to certain constraints. In
its simplest form, it is a way of assigning colours to the different vertices of a graph
such that no two adjacent vertices get the same colour; while requiring the minimum
number (referred to as the *chromatic number* of the graph) of colours. A proper
colouring requires the constraints to be strictly satisfied, and therefore none of the
adjacent nodes should get the same colour. This is called *vertex colouring*. Similarly,
edge colouring process assigns a colour to each edge so that no two adjacent edges
obtain the same colour, and *face colouring* of a planar graph assigns a colour to
each face of the graph such that no two faces that share a boundary are allotted the
same colour. Vertex colouring is the primary graph colouring problem, and colouring
problems of other types may readily be transformed into their corresponding vertex
version. For instance, the edge colouring of a graph is the same as the vertex colouring
of its line graph, and the face coloring of a planar graph is similar to the vertex coloring
of its dual.

The convention of using colours in the assignment process originates from colouring the countries of a map, where each separate geographical region is literally
coloured. This was later generalized to colouring the faces of a graph embedded in
a plane. By planar duality it became colouring the vertices, and in this form it gen-
eralizes to all graphs. In mathematical and computer representations, it is typical to
use the first few positive integers as the 'colours'. In general, one can use any finite
set as the 'set of available colours'. The nature of the colouring problem depends on
the number of colours but not on what they actually are. As shall be evident later in
this section, the neural circuits for colouring of graphs discussed in this text assign
different voltages to the nodes in the sense that the output voltages of the neurons
corresponding to various nodes are assigned different values depending upon the
actual interconnection structure of the graph.

The GCP and its variants have applications in many important tasks such as
time-tabling or event scheduling [20], register and processor allocation in digital

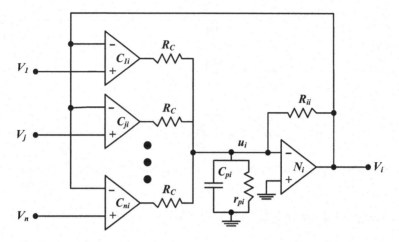

Fig. 2.11 ith neuron in the NOSYNN based circuit for graph colouring (feedback to comparators is from all neurons corresponding to the nodes which are adjacent to the ith node; $j = 1, 2, \ldots, n$; $j \neq i$)

computers [21], frequency or channel assignment in mobile communication [22] and layer assignment in VLSI design [23]. The recreational puzzle *Sudoku* can be visualized as a 'constrained-to-nine-labels' colouring on a specific graph with 81 vertices.

The NOSYNN when applied to solve GCP yielded promising results [18]. Figure 2.11 presents the ith neuron of the NOSYNN based voltage-mode neural circuit for graph colouring. The output of the ith neuron, V_i, corresponds to a voltage label (colour) assigned to the ith node. Selective connections are provided from the output of the neurons to the comparators connected in the feedback path. The voltage feedback from a node, which is connected (adjacent) to the ith node, is allowed as an input of the comparator connected in the feedback path. For non-adjacent nodes, no feedback signal arrives at the input of the ith neuron, thereby meaning that no comparators would be present for such cases. The value of R_C was chosen to be $1\,K\Omega$ and the self-feedback resistance for each neuron was calculated as

$$R_{ii} = \frac{R_C}{D} \tag{2.63}$$

where D is the maximum degree amongst all nodes in the graph to be coloured [18].

The self-feedback resistance R_{ii} can be calculated using (2.63). Next, in order to ascertain values of the resistances connected in the synaptic paths, we define a constant g_{ij} such that

$$g_{ij} = \left\{ \begin{array}{l} 1; \; ith\ node\ connected\ to\ jth\ node \\ 0; \; otherwise \end{array} \right\} \tag{2.64}$$

In Fig. 2.11, we now define x_{ij} as the voltage output of the jth comparator in the ith neuron. A mathematical expression for x_{ij} can be written as

$$x_{ij} = \frac{V_m}{2} \tanh \beta \left(V_j - V_i \right) \tag{2.65}$$

where β is the open-loop gain of the comparator (practically very high) and $\pm V_m$ are the saturation voltage levels of the comparator output.

Node equation for the inverting input terminal of the neuronal amplifier (N_i) gives the equation of motion of the ith neuron in the state space as

$$C_i \frac{du_i}{dt} = \sum_{\substack{j=1 \\ j \neq i}}^{n} \frac{x_{ij}}{R_{ij}} + \frac{V_i}{R_{ii}} - \frac{u_i}{R_i} \tag{2.66}$$

where

$$\frac{1}{R_i} = \sum_{\substack{j=1 \\ j \neq i}}^{n} \frac{1}{R_{ij}} + \frac{1}{R_{ii}} + \frac{1}{r_i} \tag{2.67}$$

Using (2.66), the NOSYNN-based graph coloring network of Fig. 2.11, which employs bipolar voltage-mode comparators, can be shown to be associated with the following energy function [18]:

$$E = \frac{1}{2} \sum_{i=1}^{n} \frac{V_i^2}{2R_{ii}} - \frac{V_m}{4\beta R_c} \sum_{i=1}^{n} \sum_{\substack{j=1 \\ j \neq i}}^{n} g_{ij} \ln \cosh \left(\beta \left(V_j - V_i \right) \right)$$

$$- \sum_{i=1}^{n} \frac{1}{R_i} \int_0^{V_i} u_i dV \tag{2.68}$$

The last term in (2.68) turns out to be small, if the open-loop gain of the operational amplifiers used to realize the neurons is high, and is usually neglected. The remaining two terms in (2.68) work in tandem by balancing each other to colour a graph properly.

2.2.3 Graph Coloring Using Modified Voltage-Mode NOSYNN

Although the NOSYNN based graph colouring network, the ith neuron of which is presented in Fig. 2.11, outperforms other existing hardware solutions for GCP [1], a minor modification in the circuit can bring about further improvements in the

Fig. 2.12 Transfer characteristics of the voltage-mode comparator modified to yield unipolar voltage outputs. **a** Case of outputs being 0 or $\pm V_m$ and **b** case of outputs being $-V_m$ or 0

Fig. 2.13 Obtaining the unipolar characteristics of Fig. 2.12 by using a bipolar comparator and a diode. **a** Circuit for obtaining the transfer characteristics of Fig. 2.12a, **b** circuit for obtaining the transfer characteristics of Fig. 2.12b

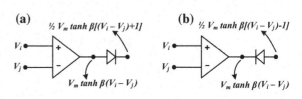

colouring performance of the network. The comparators used in the actual realization of the graph colouring NOSYNN based neural network [1] have bipolar transfer characteristics as shown in Fig. 2.6. The range of allowable voltages for the output values of various nodes is $\pm V_m$. The network then assigns voltages (traditionally called 'colours') to different nodes in the graph depending upon their adjacencies. This assignment is done by forcing the neuron to take only one out of a finite number of allowed discrete voltage levels from $[-V_m, V_m]$ for the steady state outputs [18].

However, it was observed that if the allowable range of the discrete values at the output of neurons is restricted to $[0, V_m]$ or $[-V_m, 0]$, the network can be forced to assign colours from a smaller set of available colours, thereby having the effect of reduction in the number of colours. This can be achieved by employing a unipolar comparator instead of a bipolar one as used in Fig. 2.11. The transfer characteristics of a voltage-mode unipolar comparator are presented in Fig. 2.12. As can be seen, two possibilities exist: (a) the outputs being 0 or V_m, (b) the outputs being $-V_m$ or 0. One possible circuit realization for obtaining the transfer characteristics of Fig. 2.12 is presented in Fig. 2.13, which shows a diode used in conjunction with the voltage-mode comparator yielding the characteristics as given in Fig. 2.6.

Figure 2.14 shows the ith neuron of modified voltage-mode network for graph colouring. In the proposed circuit, output voltages of different neurons represent the colors of different nodes. C_{pi} and r_{pi} denote the internal capacitance and resistance of the ith neuron respectively, u_i is the internal state and R_{ii} is the self-feedback resistance of ith neuron. The output of other neurons V_j; $(j = 1, 2 \ldots, n)$ are connected to the input of ith neuron through unipolar comparators.

The self-feedback resistance R_{ii} can be calculated using (2.63). Next, in order to ascertain values of the resistances connected in the synaptic paths, we define a constant g_{ij} such that

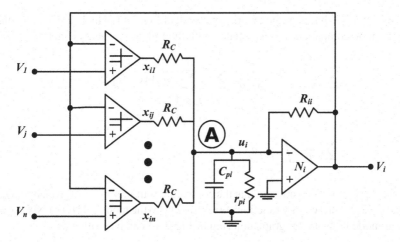

Fig. 2.14 ith neuron of the modified voltage-mode graph colouring neural network based on NOSYNN

$$g_{ij} = \left\{ \begin{array}{l} 1; \; ith \; node \; connected \; to \; jth \; node \\ 0; \; otherwise \end{array} \right\} \tag{2.69}$$

Using (2.69), the resistance in the jth synapse of the ith neuron is given by

$$R_{ij} = \frac{R_c}{g_{ij}} \tag{2.70}$$

From Fig. 2.14, x_{ij} can be written as

$$x_{ij} = \frac{V_m}{2} \left[\tanh \beta \left(V_j - V_i \right) - 1 \right] \tag{2.71}$$

where β is the open-loop gain of the comparator (practically very high) and $-V_m$ is the saturation voltage level of the comparator output.

Node equation for node 'A' gives the equation of motion of the ith neuron in the state space as

$$C_i \frac{du_i}{dt} = \sum_{\substack{j=1 \\ j \neq i}}^{n} \frac{x_{ij}}{R_{ij}} + \frac{V_i}{R_{ii}} - \frac{u_i}{R_i} \tag{2.72}$$

where

$$\frac{1}{R_i} = \sum_{\substack{j=1 \\ j \neq i}}^{n} \frac{1}{R_{ij}} + \frac{1}{R_{ii}} + \frac{1}{r_i} \tag{2.73}$$

The NOSYNN-based graph coloring network of Fig. 2.11, which employs bipolar voltage-mode comparators, is associated with the following energy function [18]:

$$
E = \frac{1}{2} \sum_{i=1}^{n} \frac{V_i^2}{2R_{ii}} - \frac{V_m}{4\beta R_c} \sum_{i=1}^{n} \sum_{\substack{j=1 \\ j \neq i}}^{n} g_{ij} \ln \cosh \left(\beta \left(V_j - V_i \right) \right)
$$

$$
- \sum_{i=1}^{n} \frac{1}{R_i} \int_{0}^{V_i} u_i dV \tag{2.74}
$$

Using (2.74), and considering that the comparator outputs are now given by (2.71) instead of (2.65), the energy function corresponding to the modified NOSYNN-based voltage-mode network for graph coloring of Fig. 2.14 can be written as

$$
E = \frac{1}{2} \sum_{i=1}^{n} \frac{V_i^2}{2R_{ii}} - \frac{V_m}{4\beta R_c} \sum_{i=1}^{n} \sum_{\substack{j=1 \\ j \neq i}}^{n} g_{ij} \ln \cosh \left(\beta \left(V_j - V_i \right) \right)
$$

$$
- \frac{V_m}{2R_c} \sum_{i=1}^{n} \sum_{\substack{j=1 \\ j \neq i}}^{n} g_{ij} V_i - \sum_{i=1}^{n} \frac{1}{R_i} \int_{0}^{V_i} u_i dV \tag{2.75}
$$

In order to prove the validity of the energy function of (2.75), the time derivative of E can be obtained as

$$
\frac{dE}{dt} = \sum_{i=1}^{n} \frac{\partial E}{\partial V_i} \frac{dv_i}{dt} = \sum_{i=1}^{n} \frac{\partial E}{\partial V_i} \frac{dV_i}{du_i} \frac{du_i}{dt} \tag{2.76}
$$

Also, for the NOSYNN-based dynamical systems, it has been shown that [18]

$$
\frac{dE}{dV_i} = C_{pi} \frac{du_i}{dt} \tag{2.77}
$$

Using (2.10) in (2.12) we get

$$
\frac{dE}{dt} = \sum_{i=1}^{n} C_i \left(\frac{du_i}{dt} \right)^2 \frac{dV_i}{du_i} \tag{2.78}
$$

The transfer characteristics of the opamp used in Fig. 2.11 implements the activation function of the neuron. With u_i being the internal state at the inverting terminal, a typical plot of the characteristics is shown in Fig. 2.15 from where it is evident that

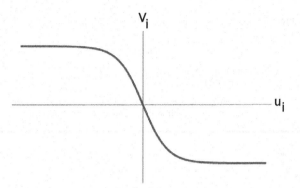

Fig. 2.15 Transfer characteristics of the opamp used to realize the neuron in Fig. 2.11

Table 2.3 Hardware implementation and PSPICE simulation test results for the proposed network

S. No.	Test Graph	Voltages of different nodes with node no.										No. of colors	Frequency of occurrence	No. of colors	Frequency of occurrence	Chromatic Number
		1	2	3	4	5	6	7	8	9	10					
1		0	$\frac{2V}{3}$	0	$\frac{2V}{3}$	0	$\frac{2V}{3}$	0	$\frac{2V}{3}$	-	-	2	25/25	2	8/10	2
														3	2/10	
2		0	V	0	V	0	V	0	V	-	-	2	22/25	2	5/10	2
		$\frac{V}{3}$	0	$\frac{V}{3}$	V	0	V	$\frac{V}{3}$	V			3	3/25	3	5/10	
3		0	V	0	$\frac{2V}{3}$	0	V	0	V	0	$\frac{2V}{3}$	3	25/25	3	10/10	2
4		0	V	0	V	0	V	0	V	0	V	2	25/25	3	10/10	2

the plot is monotonically decreasing, and therefore,

$$\frac{dV_i}{du_i} \leq 0 \tag{2.79}$$

thereby resulting in

$$\frac{dE}{dt} \leq 0 \tag{2.80}$$

with the equality being valid for

Table 2.4 Performance comparison of the proposed network with the NOSYNN-based graph coloring network of Fig. 2.11

S. No.	Test Graph	Simulation Results				Chromatic Number
		NOSYNN-based Neural Network for Graph Coloring		Modified NOSYNN-based Neural Network for Graph Coloring		
		Minimum number of colors	Average number of colors in 10 trials	Minimum number of colors	Average number of colors in 10 trials	
1		4	4.5	3	3	2
2		4	4.4	3	3	3
3		7	7.8	5	5.7	4
4		7	8.2	5	5.4	4

$$\frac{du_i}{dt} = 0 \qquad (2.81)$$

Equation (2.80) shows that the energy function can never increase with time which is one of the conditions for a valid energy function. The second criterion *viz.* the energy function must have a lower bound is also satisfied for the circuit of Fig. 2.11 wherein it may be seen that $V_1, V_2, \ldots, V_n$ are all bounded (as they are the outputs of opamps) amounting to E, as given in (2.75), having a defined lower bound.

The last term in (2.75) is usually neglected for high values of the open-loop gain of the opamp used to realize the neurons. The first term on the right hand side of (2.75) is quadratic which tries to minimize the number of colors. The second term has got a negative sign. Therefore, the energy function E will be minimized if second term is maximized. This happens when the voltages corresponding to connected nodes in a graph are far away from each other. The first two terms on the right hand side are balancing each other to color a graph properly. The third term also contributes to lowering of number of different colors by eliminating all those local minima in the energy function for which node voltages are negative. The proposed network was tested for various random graphs using PSPICE simulations as well as breadboard

Table 2.5 PSPICE simulation results for the proposed network applied to graph coloring benchmark problems

S. no.	Benchmark problem	Description (nodes, edges, chromatic number)	PSPICE simulation results	
			No. of colors	Frequency of occurrence
1	mycie13.col	11, 20, 4	5	2/10
			6	8/10
2	myciel4.col	23, 71, 5	5	3/10
			6	7/10
3	gueen5_5.col	25, 320, 5	6	6/10
			7	4/10

implementation. The results of the tests are given in Table 2.3 from which it can be seen that the proposed network gives a solution to all the problems tested and in all the cases the solution is very near to the chromatic number of the graph. The performance of the proposed network was also compared with the NOSYNN-based voltage-mode graph colouring network proposed earlier [1]. Table 2.4 presents this performance comparison. It is evident that improvements have been achieved both in the best and the average solutions for most examples. The network was further tested on three standard benchmark problems for graph coloring [24, 25]. Simulation runs of the proposed network for the benchmarking problems are presented in Table 2.5 from where it can be seen that the best solution in each case is very near to the chromatic number of the corresponding graph.

It may be mentioned that monolithic integration of the graph colouring circuit of Fig. 2.14 requires the fabrication of a large number of resistors which tends to consume a lot of chip area. A better alternative from the viewpoint of chip area conservation would be the use of transconductance elements, having voltage inputs and a current output, in the feedback paths. Conveying of the synaptic signals as currents reduces the overall number of resistances drastically and makes the circuit favorable from the viewpoint of actual VLSI implementation. Such an implementation is referred to as a mixed-mode realization and is discussed in Appendix A.

The graph colouring and ranking networks presented in [1, 19], as well as the improved graph colouring networks presented in this section, demonstrate the validity of the NOSYNN architecture for application in real-world problems. Therefore, the NOSYNN architecture has been selected as the starting point for obtaining neural circuits for other problems of interest like solving linear equations, linear programming and quadratic programming.

2.3 Chosen Problems: Description and Applications

This section deals with a description of the mathematical problems to which the various implementations of the NOSYNN are applied. These include the solution of a system of linear equations, and the minimization of linear and quadratic objective functions subject to linear constraints, referred to as Linear and Quadratic Programming Problems respectively.

2.3.1 System of Simultaneous Linear Equations

The solution of a system of linear equations, such as (2.82), has been a primary goal for computation since the time of the abacus. The earliest known references to word problems requiring the solution of a system of linear equations appear in ancient Babylonian texts dating back to *circa* 300 BC [26]. At that time, the Salamis Tablet, which was a 'counting board' and a rudimentary form of abacus, was probably used to solve linear equations. Later, around 200 BC Chinese mathematicians put forward the method of 'calculation by square tables' for solving a system of simultaneous linear equations. This method appeared in Chapter Eight of the Chinese mathematical text *Jiu Zhang Suanshu* or The Nine Chapters on the Mathematical Art [27]. Although the nomenclature is not so illuminating, the method was essentially what we now know as Gaussian Elimination. After that there was little development in methods to solve linear equations for almost two millennia. In 1750, Gabriel Cramer proposed the Cramer's rule for solving a set of n simultaneous linear equations in n variables. In 1810, Carl Friedrich Gauss devised a notation for symmetric elimination that was adopted in the 19th century by professional hand computers to solve the normal equations of least-squares problems [28]. Since then, other methods to solve a system of simultaneous linear equations that have evolved include Gauss-Jordan elimination, LU decomposition, Cholesky decomposition, Levinson recursion, etc.

A completely different approach is generally employed for large systems of simultaneous linear equations, which would otherwise take too much time or memory. The idea is to start with an initial approximation to the solution (which may not be accurate at all), and to change this chosen approximation in multiple steps to bring it nearer to the exact mathematical solution. An 'accuracy threshold' is often kept in mind while undertaking such a solution process. Once the approximation is sufficiently accurate, this is taken to be the solution to the system. Such approaches belong to the class of iterative methods for the solution of linear equations. A significant contribution from the viewpoint of solving large systems of linear equations on modern day computers came in 1966 when James Wilkinson proposed the iterative refinement method [29].

Babbage's Analytical Engine (1836) was the first attempt at automating the equation solving process. It was followed by the Atanasoff-Berry Computer (1941) which marked the transition from mechanical to an electronic computing architecture [26, 30]. However, consuming almost a minute for each addition/multiplication, the Atanasoff-Berry Computer was certainly slow. The era of fast equation solving computers was signaled by FPS-164/MAX (1984). It was followed by the Clear-Speed CSX600 (2005) which was capable of working at 25 Gflops/s consuming 10 Watts [26]. The ClearSpeed CSX600 (2005) capable of solution times of the order of microseconds is among the fastest available linear equation solvers. However, it needs to run a host operating system and is prone to "OS jitter" at high operating speeds. These issues together with the fact that its power consumption is around 10 Watts make the CSX600 unsuitable for real-time and/or portable applications. Among the current state of the art solutions, the recently developed concurrent multiprocessor based architectures for solving linear equations; only systolic/wavefront arrays [31] and the Block Data Parallel Architecture (BDPA) are suited for solving computationally intensive problems [32]. However, even with the extensive simulation tools developed for the BDPA, there is still a fundamental need to show the prospective power of such architectures. Such arrangements may not be suitable for real-time and/or portable applications where a dedicated, compact, low-power solution is desirable. This has led to research efforts being directed towards the development of specialized hardware for the solution of linear equations.

Another approach to solving linear equations was to use specialized machines. One such mechanical linear equation solver was constructed in 1936 at the Massachusetts Institute of Technology [33]. The late 1940s witnessed the advent of electronic methods for solving linear equations [34]. Several more analog circuits were reported in the next decade [35–37]. In 1988, an analog resistive network was proposed for solving linear equations [38]. Later, neural networks promising massively parallel processing and fast convergence were applied to solve linear equations [39–47]. More recently, Field Programmable Gate Arrays (FPGAs) with their inherent capability to be used as multi-million-gate system-on-chip have been employed to parallelize the task of solving large systems of equations [48].

A general system of m linear equations with n unknowns can be written as

$$
\begin{aligned}
a_{11}x_1 + a_{12}x_n + \cdots + a_{1n}x_n &= b_1 \\
a_{21}x_1 + a_{22}x_2 + \cdots + a_{2n}x_n &= b_2 \\
&\vdots \\
a_{m1}x_1 + a_{m2}x_n + \cdots + a_{mn}x_n &= b_m
\end{aligned}
\tag{2.82}
$$

where $x_1, x_2, \ldots, x_n$ are the unknowns, $a_{11}, a_{12}, \ldots, a_{mn}$ are the coefficients of the system of linear equations, and $b_1, b_2, \ldots, b_m$ are the constant terms. Often the coefficients and unknowns are real or complex numbers, but polynomials and elements of an abstract algebraic structure are also possible. In general, for the set of Eqs. (2.82) to possess a unique solution, the linear equations must be linearly independent and $m = n$.

Solving a system of simultaneous linear equations is one of the most fundamental problems in algebra. In the context of modern day systems, solving linear equations is an integral part of many scientific and engineering problems *viz.* curve fitting, electrical circuit analysis, multiple correlation as well as real time applications like real-time speech coding, image processing, stochastic modeling, and computer-aided realistic three-dimensional image synthesis [26, 32, 49].

As mentioned above, present day systems rely on software algorithms to arrive at the solution for a given set of linear equations. However computer algorithms, by virtue of their sequential nature, tend to have a solution time which may not be practical for real-time systems. Further, running of an algorithm may also require an operating system which may not be feasible on a portable system where a dedicated, compact, low-power solution is desirable. This has led to research efforts being directed towards the development of specialized hardware for the solution of linear equations. The material presented here is an attempt to develop voltage-mode and mixed-mode circuits for solving linear equations. Further, it has been shown that minor modifications in the proposed neural networks for solving linear equations enable them to solve two important constrained optimization problems *viz.* linear and quadratic programming [49].

2.3.2 Linear Programming Problem

Mathematical programming, in general, is concerned with the determination of a minimum or a maximum of a function of several variables, which are required to satisfy a number of constraints. Such solutions are sought in diverse fields including engineering, operations research, management science, computer science, numerical analysis, and economics [49, 50].

A general mathematical programming problem can be stated as [50]:

$$Minimize \quad f(x)$$

subject to

$$g_i\,(\mathbf{x}) \geq 0 \ (i = 1, 2, \ldots, m) \tag{2.83}$$

$$h_j\,(\mathbf{x}) = 0 \ (j = 1, 2, \ldots, p) \tag{2.84}$$

$$\mathbf{x} \in S \tag{2.85}$$

where $x = (x_1, x_2, \ldots, x_n)^T$ is the vector of unknown decision variables, and $f, g_i(i = 1, 2, \ldots, m), h_j(j = 1, 2, \ldots, p)$ are the real-valued functions of the n real variables $x_1, x_2, \ldots, x_n$.

In this formulation, the function f is called the *objective function*, and inequalities (2.83), Eqs. (2.84) and the set restrictions (2.85) are referred to as the *constraints*. It may be mentioned that although the mathematical programming problem (MPP) has been stated as a minimization problem in the description above, the same may readily be converted into a maximization problem without any loss of generality, by using the identity

$$\max f(x) = -\min\,[-f(x)] \tag{2.86}$$

As a special case, if all the functions appearing in the MPP are linear in the decision variables x, the problem is referred to as a *linear programming problem* (LPP). Such LPPs have been investigated extensively over the past decades, in view of their fundamental roles arising in a wide variety of engineering and scientific applications, such as, pattern recognition [51], signal processing [52], human movement analysis [53], robotic control [54], and data regression [55]. Other real life applications include portfolio optimization [56], crew scheduling [57], manufacturing and transportation [58], telecommunications [59], and the TSP [60].

Numerous software algorithms are available for the solution of LPP including the hugely popular Simplex algorithm, Criss-cross algorithm, Conic sampling algorithm, Ellipsoid algorithm and the Projective algorithm. A multitude of software packages, both open source and proprietary are available for linear programming. Examples include:

- **OpenOpt.** It is a universal cross-platform numerical optimization framework.
- **Cassowary constraint solver.** It is an incremental constraint solving package that efficiently solves systems of linear equalities and inequalities.
- **LiPS.** A free optimization package intended for solving linear, integer and goal programming problems.
- **SuanShu.** A Java-based math library that supports linear programming and other kinds of numerical optimization.

2.3.3 Quadratic Programming Problem

A *quadratic programming problem* (QPP) is a special case of the general MPP discussed in the previous section for which objective function f is a second-order function of the decision variables while the constraints remain linear [50]. Moreover, optimization problems with nonlinear objective functions are usually approximated by a second-order system and then solved by standard quadratic programming techniques. Quadratic programming problems arise naturally in a variety of applications, such as structural analysis [61], optimal control [62], plastic analysis [63], antenna array pattern synthesis [64], geometric optimization [65], propulsion physics [66], multi-commodity networks [67], etc.

A variety of methods are available for solving commonly occurring QPPs *viz.* interior point method, active set method, augmented Lagrangian technique, conjugate gradient method, gradient projection technique, and even extensions of the simplex algorithm are available for QPPs. Currently popular methods for solving QPPs include

- **Gurobi Solver.** It is a software tool with parallel algorithms for large-scale linear programs, quadratic programs and mixed-integer programs. Since it is free to use this program for academic purposes, it is quite popular among researchers working in the area.

- **OPTI Toolbox.** It is a free toolbox for the very popular MATLAB package for solving linear, nonlinear, continuous and discrete optimization problems.
- **NAG Numerical Library.** It is a collection of mathematical and statistical routines developed by the *Numerical Algorithms Group* for multiple programming languages (C, C++, Fortran, Visual Basic, Java and C♯) and also for stand-alone engineering software packages (MATLAB, Excel, LabVIEW). The NAG Library includes routines for QPP with both sparse and non-sparse linear constraint matrices, along with routines for the optimization of linear and nonlinear functions with nonlinear, bounded or no constraints.

2.4 Overview of Relevant Literature

Since the re-emergence of NNs as a viable alternative for the solution of combinatorial optimization problems, several neural networks have been proposed to solve systems of linear equations, LPP and QPP. Jang, Lee and Shin proposed a neural network for matrix inversion which can be applied, with some modifications, to solve simultaneous linear equations [39]. Finding the inverse of a matrix was expressed as an optimization problem and the dissipative dynamics approach of Hopfield architecture was employed to design the electronic hardware. For an $n \times n$ matrix, this method uses n essentially similar networks, each network optimizing an energy function. Chakraborty et al. improved upon the network of [39] and employed only a single energy function having a global minimum at a solution, if the chosen system of linear equations is solvable [40]. However, with a convergence time to the order of 10 s obtained during the course of simulations, their network was certainly slow.

Around the same time, Wang proposed an electronic realization of a recurrent neural network for solving linear equations [41]. To solve an n-variable system of equations, the network employs three operational amplifiers, one capacitor and $(n+5)$ resistances to emulate a single neuron and the time to arrive at the solution is of the order of hundreds of milliseconds [41]. Cichocki and Unbehauen presented a neural circuit for solving linear equations which was able to provide a significantly improved solution time of the order of microseconds but at the cost of increased hardware complexity [42]. Each neuron in [42] comprises of three weighted summers and an inverting integrator. Wang and Li employed a linear activation function in their neural network for solving linear equations [44]. The electronic realization of the network presented in [44] is very similar to the one proposed in [41]. Zhang et al. employed Anti-Hebbian synapses for the solution of linear equations [45]. Xia et al. came up with a linear equation solver [46], which is essentially a generalized neural network based implementation of Censor and Elfving's method for linear inequalities [68]. They used an approach similar to [41], utilizing weighted adders and integrators to realize the neurons. More recently, Jiang proposed a recurrent neural network for the on-line solution of linear equations with time varying variables [47].

LPP has also received considerable research attention from the neural networks community. The first solution of the linear programming problem was proposed by

Tank and Hopfield wherein they used the continuous-time Hopfield network [4]. From the computational aspect, the operation of Hopfield network for an optimization problem, like the LPP, manages a dynamic system characterized by an energy function, which is the combination of the objective function and the constraints of the original problem [69]. However, the Hopfield network when applied to solve LPP fails to satisfy the Kuhn-Tucker optimality conditions for a minimizer. Over the years, the penalty function approach has become a popular technique for solving optimization problems. Kennedy and Chua proposed an improved version of Tank and Hopfield's network for LPP in which an inexact penalty function was considered [70]. The requirement of setting a large number of parameters was a major drawback of Kennedy and Chua's LPP network [52]. Also, only approximate solutions were obtained as for true minimizations, the penalty parameter was required to be infinitely high which was impossible realistically. Rodriguez-Vazquez et al. used a different penalty method to transform the given LPP into an unconstrained optimization problem [71]. For solving a LPP in n variables with m constraints, their network employed n integrators, m summers, a *constraint block* comprising of m comparators and $(m + 1)$ AND gates. Although Rodriguez-Vazquez et al. later pointed out that their network had no equilibrium point in the classical sense [72], investigations by Lan et al. proved that the network can indeed converge to an optimal solution of the given problem from any arbitrary initial condition [73]. Maa and Shanblatt employed a two-phase neural network architecture for solving LPPs [74] but the network was certainly slow being able to provide solutions in times of the order of *seconds*. Chong et al. analyzed a class of neural network models for the solution of LPPs by dynamic gradient approaches based on exact non-differentiable penalty functions [75]. They also developed an analytical tool aimed at helping the system converge to a solution within a finite time. In the sample LPP solved in [75], the time taken by the network to arrive at the solution was around 600 ms. In an approach different from the penalty function methods, Zhu et al. proposed a Lagrange method for solving LPPs through Hopfield networks by employing two distinct types of neurons *viz.* the *variable* neurons and the *Langrangian* neurons [76]. Instead of following a direct descent approach of the penalty function, the network searched for a first-order necessary condition of optimality in the state space. Xia and Wang used bounded variables to construct a new neural network approach to solve LPP with no penalty parameters. They suggested that the equilibrium point is the same as the exact solution when the primal and dual problems are solved simultaneously [77]. However, only a block arrangement was provided in [77], and no actual implementation was suggested. More recently, Malek and Yari proposed two new methods for solving the LPP and presented optimal solutions with efficient convergence within a finite time [78]. Lastly, Ghasabi-Oskoei et al. have presented a recurrent neural network model for solving LPP based on a dynamical system using arbitrary initial conditions. The method does not require analog multipliers thereby reducing the system complexity [79]. However, in this case too, only a block arrangement is presented without details of the actual hardware realization. Furthermore, the solution time for the network is in seconds thereby making it unsuitable for real-time applications.

Various methods to solve QPP by employing neural network approaches are available in the technical literature. Kennedy and Chua extended the Tank and Hopfield network by developing a neural network for solving nonlinear programming problems, by satisfaction of the Karush–Kuhn–Tucker optimality conditions [70]. However, the need to set a penalty parameter means that the network can generate approximate solutions only and implementation problems arise when the penalty parameter is large. Each variable amplifier comprises of two opamps, two resistors and one capacitor whereas for satisfying each constraint, the constraint amplifier employs three opamps, two resistors and one diode [70]. Wang proposed a recurrent neural network for solving QPPs with equality constraints. The network is asymptotically stable and is able to generate optimal solutions to quadratic programs with equality constraints. An opamp based circuit realization of the network is also presented which requires $(n + m)$ neurons for solving a QPP in n variables with m constraints. Each neuron is made up of a summer, an integrator, and an inverter consuming three opamps, one capacitor and $(n + 5)$ resistors [80]. Wang's network is not suitable for real time applications as it takes around 50 ms to arrive at the solution [80]. A rigorous analysis of the prominent neural networks for QPP, available till that time (1992), is presented in [74]. Forti and Tesi presented new conditions capable of ensuring existence, uniqueness, and global asymptotic stability of the equilibrium point for Kennedy and Chua's network [81]. Wu et al. proposed two neural network models for solving LPP and QPP, the convergence of which was not dependent on the network parameters [82]. Around the same time, Xia also put forward a neural network capable of solving both LPP and QPP in which no parameter tuning was necessary. Moreover, the actual hardware implementation was somewhat simplified, as compared to its contemporaries, because of the fact that no analog multipliers were required for the variables [83]. To solve a QPP in n variables with m constraints, Xia's network consisted of $(2m^2 + 4mn)$ amplifiers, $(2m^2 + 4mn + 3m + 3)$ summers, $(n + m)$ integrators, and n limiters. Tao et al. further simplified the network of Xia [83], and reduced the system complexity [84]. More recently, Liu and Wang presented a one layer feedback neural network with a discontinuous hard-limiting activation function for solving QPP in which the number of neurons is the same as the number of decision variables [85]. Each neuron in [85] is composed of 2 adders, $(3n + 1)$ resistors, one limiter and one intergrator. Although significant reduction in circuit complexity is achieved, the time that the circuit takes to arrive at the correct solution is of the order of *seconds* thereby making the circuit unsuitable for applications requiring fast solution times. A comprehensive bibliography of the technical literature related to QPP is also available on the world wide web [86].

2.5 Summary

A brief overview of the background information which is pertinent to the content of subsequent chapters is presented. The popular Hopfield network is first explained and issues in convergence and other limitations associated with the network are highlighted. The NOSYNN, which was proposed as an alternative to the Hopfield network, is then explained and applications of the NOSYNN to graph colouring and sorting problems are also elaborated. Thereafter, the chosen problems for which the hardware implementations of the NOSYNN have been tested, *viz.* the solution of a system of simultaneous linear equations and linear & quadratic programming, are discussed. A survey of existing neural networks for the solution of the chosen problems is then presented.

From this point onwards, the book presents neural architectures for the solution of the chosen problems starting with exploring the possibility of solving linear equations using the Hopfield network.

References

1. Rahman, S.A., Jayadeva, C., Dutta Roy, S.C.: Neural network approach to graph colouring. Electron. Lett. **35**(14), 1173–1175 (1999)
2. Hopfield, J.J.: Neural networks and physical systems with emergent collective computational abilities. Proc. Natl. Acad. Sci. **79**(8), 2554–2558 (1982)
3. Hopfield, J.J., Tank, D.W.: "Neural" computation of decisions in optimization problems. Biol. Cybernet. **52**, 141–152 (1985)
4. Tank, D., Hopfield, J.: Simple 'neural' optimization networks: an A/D converter, signal decision circuit, and a linear programming circuit. IEEE Trans. Circ. Syst. **33**(5), 533–541 (1986)
5. Hopfield, J.J.: Neurons with graded response have collective computational properties like those of two-state neurons. Proc. Natl. Acad. Sci. **81**(10), 3088–3092 (1984)
6. Vidyasagar, M.: Location and stability of the high-gain equilibria of nonlinear neural networks. IEEE Trans. Neural Netw. **4**(4), 660–672 (1993)
7. Wilson, G.V., Pawley, G.S.: On the stability of the travelling salesman problem algorithm of hopfield and tank. Biol. Cybern. **58**(1), 63–70 (1988)
8. Kamgar-Parsi, B., Kamgar-Parsi, B.: Dynamical stability and parameter selection in neural optimization. In Proceedings of International Joint Conference on Neural Networks (IJCNN), pp. 566–571, Baltimore, USA (1992)
9. Aiyer, S.V.B., Niranjan, M., Fallside, F.: A theoretical investigation into the performance of the hopfield model. IEEE Trans. Neural Netw. **1**(2), 204–215 (1990)
10. Gee, A.H.: Problem Solving with Optimization Networks. PhD thesis, University of Cambridge, UK (1993)
11. Van den Bout, D.E., Miller, T.K.: A traveling salesman objective function that works. In: Proceedings of IEEE International Conference on Neural Networks, pp. 299–303, San Diego, USA (1988)
12. Nonaka, H., Kobayashi, Y.: Sub-optimal solution screening in optimization by neural networks. In: Proceedings of International Joint Conference on Neural Networks (IJCNN), pp. 606–611, Baltimore, USA (1992)
13. Foo, Y.P.S., Szu, H.: Solving large-scale optimization problems by divide-and-conquer neural networks. In: Proceedings of International Joint Conference on Neural Networks (IJCNN), pp. 507–511, Washington, DC, USA (1989)
14. Lo, J.T.-H.: A new approach to global optimization and its applications to neural networks. In: Proceedings of International Joint Conference on Neural Networks (IJCNN), pp. 600–605, Baltimore, USA (1992)
15. Amartur, S.C., Piraino, D., Takefuji, Y.: Optimization neural networks for the segmentation of magnetic resonance images. IEEE Trans. Med. Imaging **11**(2), 215–220 (1992)

16. Chen, L., Aihara, K.: Chaotic simulated annealing by a neural network model with transient chaos. Neural Netw. **8**(6), 915–930 (1995)
17. Arabas, J., Kozdrowski, S.: Applying an evolutionary algorithm to telecommunication network design. IEEE Trans. Evol. Comput. **5**(4), 309–322 (2001)
18. Rahman, S.A.: A nonlinear synapse neural network and its applications. PhD thesis, Department of Electrical Engineering, Indian Institute of Technology, Delhi, India (2007)
19. Jayadeva, Rahman, S.A.: A neural network with O(N) neurons for ranking N numbers in O(1/N) time. IEEE Trans. Circ. Syst. I: Regular Papers **51**(10):2044–2051 (2004)
20. Jensen, T.R., Toft, B.: Graph Coloring Problems. John Wiley & Sons, New York (1994)
21. Gassen, D.W., Carothers, J.D.: Graph color minimization using neural networks. In: Proceedings of International Joint Conference on Neural Networks (IJCNN), pp. 1541–1544, Nagoya, Japan (1993)
22. El-Fishawy, N.A., Hadhood, M.M., Elnoubi, S., EL-Sersy, W.: A modified hopfield neural network algorithm for cellular radio channel assignment. In: Proceedings of TENCON, pp. 213–216, Kuala Lumpur, Malaysia (2000)
23. LaPaugh, A.S.: VLSI Layout algorithms. Chapman & Hall/CRC, Boca Raton (2010)
24. Yue, T.-W., Lee, Z.Z.: A Q'tron neural-network approach to solve the graph coloring problems. In: Proceedings of 19th IEEE International Conference on Tools with Artificial Intelligence (ICTAI), pp. 19–23, Paris, France (2007)
25. Graph Coloring Instances. http://mat.gsia.cmu.edu/COLOR/instances.html. Accessed 10 February 2012
26. Gustafson, J.L.: The quest for linear equation solvers and the invention of electronic digital computing. In: Proceedings of IEEE John Vincent Atanasoff 2006 International Symposium on Modern Computing (JVA'06), pp. 10–16, Sofia (2006)
27. Yong, L.L.: Jiu Zhang Suanshu (nine chapters on the mathematical art): an overview. Arch. Hist. Exact Sci. **47**(1), 1–51 (1994)
28. Kleiner, I.: History of Linear Algebra, pp. 79–89. Birkhauser, Basel (2007)
29. Peters, G., Wilkinson, J.H., Martin, R.S.: Iterative refinement of the solution of a positive definite system of equations. Numer. Math. **8**, 203–216 (1966)
30. White, S.: A brief history of computing—complete timeline. http://trillian.randomstuff.org.uk/stephen/history/timeline.html. Accessed 30 October 2012
31. Kung, S.: VLSI array processors. IEEE ASSP Mag. **2**(3), 4–22 (1985)
32. Wilburn, V.C., Ko, H.-L., Alexander, W.E.: An algorithm and architecture for the parallel solution of systems of linear equations. In: Proceedings of IEEE Fifteenth Annual International Phoenix Conference on Computer and Communications, pp. 392–398, Scottsdale, USA, March 1996
33. Wilbur, J.B.: The mechanical solution of simultaneous equations. J. Franklin Inst. **222**, 715–724 (1936)
34. Walker, R.M.: An analogue computer for the solution of linear simultaneous equations. Proc. IRE Waves Electrons Section **37**(12), 1467–1473 (1949)
35. Ackerman, S.: Precise solutions of linear simultaneous equations using a low cost analog. Rev. Sci. Instrum. **22**(10), 746–748 (1951)
36. Many, A., Oppenheim, U., Amitsur, S.: An electrical computer for the solution of linear simultaneous equations. Rev. Sci. Instrum. **24**(2), 112–116 (1953)
37. Mitra, S.K.: Electrical analog computing machine for solving linear equations and related problems. Rev. Sci. Instrum. **26**(5), 453–457 (1955)
38. Hutchinson, J., Koch, C., Luo, J., Mead, C.: Computing motion using analog and binary resistive networks. Computer **21**(3), 52–63 (1988)
39. Jang, J., Lee, S., Shin, S.: An Optimization Network for Matrix Inversion, pp. 397–401. American Institute of Physics, New York (1988)
40. Chakraborty, K., Mehrotra, K., Mohan, C.K., Ranka, S.: An optimization network for solving a set of simultaneous linear equations. In: Proceedings of International Joint Conference on Neural Networks (IJCNN), pp. 516–521, Baltimore, USA, June 1992

41. Wang, J.: Electronic realization of recurrent neural network for solving simultaneous linear equations. Electron. Lett. **28**(5), 493–495 (1992)
42. Cichocki, A., Unbehauen, R.: Neural networks for solving systems of linear equations and related problems. IEEE Trans. Circ. Syst. I: Fundam. Theory Appl. **39**(2), 124–138 (1992)
43. Cichocki, A., Unbehauen, R.: Neural networks for solving systems of linear equations. Part ii. Minimax and least absolute value problems. IEEE Trans. Circ. Syst. II: Analog Digital Signal Processing **39**(9), 619–633 (1992)
44. Wang, J., Li, H.: Solving simultaneous linear equations using recurrent neural networks. Inf. Sci. Intell. Syst. **76**(3), 255–277 (1994)
45. Zhang, K., Ganis, G., Sereno, M.I.: Anti-hebbian synapses as a linear equation solver. In: Proceedings of International Conference on Neural Networks, pp. 387–389, Houston, TX, USA, June 1997
46. Xia, Y., Wang, J., Hung, D.L.: Recurrent neural networks for solving linear inequalities and equations. IEEE Trans. Circ. Syst. I: Fundam. Theory Appl. **46**(4), 452–462 (1999)
47. Jiang, D.: Analog computing for real-time solution of time-varying linear equations. In: Proceedings of International Conference on Communications, Circuits and Systems (ICCCAS), pp. 1367–1371, June 2004
48. Wang, X., Ziavras, S.G.: Parallel direct solution of linear equations on FPGA-based machines. In: Proceedings of International Parallel and Distributed Processing Symposium (IPDPS'03), pp. 113–120, Nice, France, April 2003
49. Kreyszig, E.: Advanced Engineering Mathematics, 8th edn. Wiley-India (2006)
50. Kambo, N.S.: Mathematical Programming Techniques, revised edn. Affilated East-West Press Pvt Ltd., New Delhi (1991)
51. Anguita, D., Boni, A., Ridella, S.: A digital architecture for support vector machines: theory, algorithm, and FPGA implementation. IEEE Trans. Neural Netw. **14**(5), 993–1009 (2003)
52. Cichocki, A., Unbehauen, R.: Neural Networks for Optimization and Signal Processing. Wiley, Chichester (1993)
53. Iqbal, K., Pai, Y.C.: Predicted region of stability for balance recovery: motion at the knee joint can improve termination of forward movement. J. Biomech. **13**(12), 1619–1627 (2000)
54. Zhang, Y.: Towards piecewise-linear primal neural networks for optimization and redundant robotics. In: Proceedings of IEEE International Conference on Networking, Sensing and Control, pp. 374–379, Fort Lauderdale, Florida, USA (2006)
55. Zhang, Y., Leithead, W.E.: Exploiting hessian matrix and trust-region algorithm in hyperparameters estimation of gaussian process. Appl. Math. Comput. **171**(2), 1264–1281 (2005)
56. Young, M.R.: A minimax portfolio selection rule with linear programming solution. Manage. Sci. **44**(5), 673–683 (1988)
57. Bixby, R.E., Gregory, J.W., Lustig, I.J., Marsten, R.E., Shanno, D.F.: Very large-scale linear programming: a case study in combining interior point and simplex methods. Oper. Res. **40**(5), 885–897 (1992)
58. Peidro, D., Mula, J., Jimenez, M., del Mar Botella, M.: A fuzzy linear programming based approach for tactical supply chain planning in an uncertainty environment. Eur. J. Oper. Res. **205**(16), 65–80 (2010)
59. Chertkov, M., Stepanov, M.G.: An efficient pseudocodeword search algorithm for linear programming decoding of LDPC codes. IEEE Trans. Inf. Theory **54**(4), 1514–1520 (2008)
60. ChvÃtal, V., Cook, W., Dantzig, G.B., Fulkerson, D.R., Johnson, S.M.: Solution of a Large-Scale Traveling-Salesman Problem, pp. 7–28. Springer, Berlin, (2010)
61. Atkociunas, J.: Quadratic programming for degenerate shakedown problems of bar structures. Mech. Res. Commun. **23**(2), 195–206 (1996)
62. Bartlett, R.A., Wachter, A., Biegler, L.T.: Active set vs. interior point strategies for model predictive control. In: Proceedings of American Control Conference, pp. 4229–4233, Chicago, USA, June 2000
63. Maier, G., Munro, J.: Mathematical programming applications to engineering plastic analysis. Appl. Mech. Rev. **35**, 1631–1643 (1982)

64. Nordebo, S., Zang, Z., Claesson, I.: A semi-infinite quadratic programming algorithm with applications to array pattern synthesis. IEEE Trans. Circ. Syst. II: Analog and Digital Signal Processing **48**(3), 225–232 (2001)
65. Schonherr, S.: Quadratic programming in geometric optimization: theory, implementation, and applications. PhD thesis, Swiss Federal Institute of Technology, Zurich (2002)
66. Borguet, S., Leonard, O.: A quadratic programming framework for constrained and robust jet engine health monitoring. Progr. Propul. Phys. **1**, 669–692 (2009)
67. Dembo, R.S., Tulowitzki, U.: Computing equilibria on large multicommodity networks: an application of truncated quadratic programming algorithms. Networks **18**(4), 273–284 (1988)
68. Censor, Y., Elfving, T.: New methods for linear inequalities. Linear Algebra Appl. **42**, 199–211 (1982)
69. Wen, U.-P., Lan, K.-M., Shih, H.-S.: A review of hopfield neural networks for solving mathematical programming problems. Eur. J. Oper. Res. **198**(3), 675–687 (2009)
70. Kennedy, M.P., Chua, L.O.: Neural networks for nonlinear programming. IEEE Trans. Circ. Syst. **35**(5), 554–562 (1988)
71. Rodriguez-Vazquez, A., Rueda, A., Huertas, J.L., Dominguez-Castro, R.: Switched-capacitor neural networks for linear programming. Electron. Lett. **24**(8), 496–498 (1988)
72. Rodriguez-Vazquez, A., Dominguez-Castro, R., Rueda, A., Huertas, J.L., Sanchez-Sinencio, E.: Nonlinear switched capacitor 'neural' networks for optimization problems. IEEE Trans. Circ. Syst. **37**(3), 384–398 (1990)
73. Lan, K.-M., Wen, U.-P., Shih, H.-S., Lee, E.S.: A hybrid neural network approach to bilevel programming problems. Appl. Math. Lett. **20**(8), 880–884 (2007)
74. Maa, C.-Y., Shanblatt, M.A.: Linear and quadratic programming neural network analysis. IEEE Trans. Neural Netw. **3**(4), 580–594 (1992)
75. Chong, E.K.P., Hui, S., Zak, S.H.: An analysis of a class of neural networks for solving linear programming problems. IEEE Trans. Autom. Control **44**(11), 1995–2006 (1999)
76. Zhu, X., Zhang, S., Constantinides, A.G.: Lagrange neural networks for linear programming. J. Parallel Distrib. Comput. **14**(3), 354–360 (1992)
77. Xia, Y., Wang, J.: Neural network for solving linear programming problems with bounded variables. IEEE Trans. Neural Netw. **6**(2), 515–519 (1995)
78. Malek, A., Yari, A.: Primal-dual solution for the linear programming problems using neural networks. Appl. Math. Comput. **167**(1), 198–211 (2005)
79. Ghasabi-Oskoei, H., Malek, A., Ahmadi, A.: Novel artificial neural network with simulation aspects for solving linear and quadratic programming problems. Comput. Math. Appl. **53**(9), 1439–1454 (2007)
80. Wang, J.: Recurrent neural network for solving quadratic programming problems with equality constraints. Electron. Lett. **28**(14), 1345–1347 (1992)
81. Forti, M., Tesi, A.: New conditions for global stability of neural networks with application to linear and quadratic programming problems. IEEE Trans. Circ. Syst. I: Fundam. Theory Appl. **42**(7), 354–366 (1995)
82. Wu, X.-Y., Xia, Y.-S., Li, J., Chen, W.-K.: A high-performance neural network for solving linear and quadratic programming problems. IEEE Trans. Neural Netw. **7**(3), 643–651 (1996)
83. Xia, Y.: A new neural network for solving linear and quadratic programming problems. IEEE Trans. Neural Netw. **7**(6), 1544–1548 (1996)
84. Tao, Q., Cao, J., Sun, D.: A simple and high performance neural network for quadratic programming problems. Appl. Math. Comput. **124**(2), 251–260 (2001)
85. Liu, Q., Wang, J.: A one-layer recurrent neural network with a discontinuous hard-limiting activation function for quadratic programming. IEEE Trans. Neural Netw. **19**(4), 558–570 (2008)
86. Gould, N.I.M., Toint, P.L.: A quadratic programming bibliography. Technical report, RAL Numerical Analysis Group, March 2010

Chapter 3
Voltage-Mode Neural Network for the Solution of Linear Equations

3.1 Introduction

The application of Non-Linear Synapse Neural Network (NOSYNN) for the solution of sorting and ranking has already been discussed in Chap. 2. In the present chapter, it is demonstrated that NOSYNN-based architectures are also more adept at the task of solving a system of simultaneous linear equations, while enjoying fast convergence and reduced circuit complexity in comparison with other neural network hardware.

This chapter is organized as follows. Section 3.2 discusses the pros and cons of employing the HNN for the task of solving linear equations. It is shown that the HNN in its standard form is not suitable for the solution of linear equations. It is further shown that even with suitable modifications, the HNN solves only trivial problems, and therefore alternative structures are needed. Section 3.3 contains the details of the NOSYNN-based network along with the design equations. Proof of the validity of the energy function is provided. In the same section, it is also shown that the stable states of the neurons in the NOSYNN-based network, which are governed by the stationary point (minimum) of the energy function, correspond exactly with the solution point of the system of linear equations. The convergence time for the neuron outputs is discussed too. Section 3.4 presents the results of PSPICE simulation of the circuit applied to solve various sample equation sets. Results of hardware implementation of the network for small-sized problems are contained in Sect. 3.5. A low-voltage CMOS-compatible implementation of the NOSYNN-based linear equation solver is presented in Sect. 3.6. Section 3.7 presents a comparison of the NOSYNN-based linear equation solver with other neural network available for the purpose. A discussion on VLSI implementability of the circuit appears in Sect. 3.8. Some conclusive remarks appear in Sect. 3.9.

M. S. Ansari, *Non-Linear Feedback Neural Networks*,   55
Studies in Computational Intelligence 508, DOI: 10.1007/978-81-322-1563-9_3,
© Springer India 2014

3.2 Solving Linear Equations Using the Hopfield Neural Network

As discussed in Chap. 2, the Hopfield network is known to be one of the most influential and popular models in neural systems research [1]. Many other models that appeared in the technical literature after the advent of Hopfield networks, like the bidirectional associative memories [2], Boltzmann machines [3], Q-state attractor neural networks [4], etc., are either direct variants or generalizations of the Hopfield network [1]. The Hopfield neural network has been put to use in a variety of applications, including but not limited to, Content Addressable Memory (CAM), solution of Travelling Salesman Problem (TSP), analog-to-digital conversion, signal decision and linear programming. To embed and solve problems related to optimization, neural control and signal processing, the Hopfield neural network has to be designed to possess only one and globally stable equilibrium point, so as to avoid the risk of spurious responses arising due to the common problem of local minima [5]. In order to achieve these goals, it is customary to impose constraint conditions on the interconnection matrix of the system [6]. For instance, most of the study and application of the Hopfield-type neural networks has been carried out with the assumption that the interconnection weight matrix ($\mathbf{W}$ in (2.2)) is symmetric [1].

This section explores the possibility of solving a system of simultaneous linear equations using Hopfield neural network based approaches. Section 3.2.1 presents an overview of Hopfield's original network along with a discussion on why the standard Hopfield network is not suitable for the task of solving linear equations. Section 3.2.2 presents a modified Hopfield neural network applied to solve linear equations for the case when the coefficient matrix, $\mathbf{A}$ ($=(a_{ij})_{n \times n}$ in (2.82)) corresponding to the system of linear equations is *symmetric*. PSPICE simulation results for various systems of simultaneous linear equations with symmetric coefficient matrices are also presented.

3.2.1 The Hopfield Network

A brief overview of the Hopfield network has already been presented in Chap. 2 wherein the differential equation governing the behaviour of the ith neuron in the network (reproduced in Fig. 3.1) was given as

$$C_i \frac{du_i}{dt} = \sum_{j=1}^{n} W_{ij} v_j - \frac{u_i}{R_i} + i_i$$

$$v_i = g_i(u_i), i = 1, 2, \ldots, n \qquad (3.1)$$

where $C_i \geq 0$, $R_i \geq 0$, and i_i are the capacity, resistance, and bias, respectively, and u_i and v_i are the input and output of the ith neuron respectively and $g_i(.)$ is

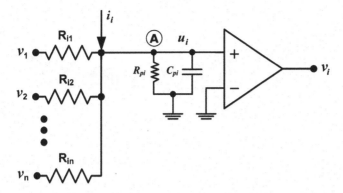

Fig. 3.1 ith neuron of Hopfield neural network

the characteristic of the ith neuron [7–9]. W_{ij} are the elements of the weight matrix **W** and these weights are implemented using resistances $R_{ij}(= 1/W_{ij})$ in the electronic realization of the Hopfield network as shown in Fig. 3.1. Hopfield studied such networks under the assumptions that

$$W_{ii} = 0 \text{ for all } i \tag{3.2}$$

and

$$W_{ij} = W_{ji} \text{ for all } i, j \tag{3.3}$$

While the assumption of (3.2) implies that no self-interactions (self-feedbacks) are allowed for the neurons in the original Hopfield network, the assumption of (3.3) puts a more stringent condition on the interactions between different neurons wherein only symmetric interactions are allowed. This means that the coefficient weight matrix **W** $(=(W_{ij})_{n \times n})$ is assumed to be symmetric. For such cases, where there are no self-interactions and a symmetric interconnection matrix exists, the Hopfield network is totally stable [10]. Under such conditions, the energy function corresponding to the Hopfield network, the ith neuron of which is shown in Fig. 3.1, has already been derived in (2.11) which is reproduced here for reference (3.4).

$$E = -\frac{1}{2} \sum_i \sum_j W_{ij} v_i v_j - \sum_i i_i v_i + \sum_i \frac{1}{R_i} \int_0^{v_i} g_i^{-1}(\mathbf{v}) d\mathbf{v} \tag{3.4}$$

The last term in (3.4) is only significant near the saturating values of the opamp and is usually neglected [11]. A plot showing a typical energy function for a 2–neuron system with a symmetric weight matrix with zero diagonal elements (fulfilling the requirements of (3.2) and (3.3)), and no external bias i_i, is shown in Fig. 3.2. It is evident that the stable states for the plot shown in Fig. 3.2 lie at the corners of the hypercube $[-V_m, V_m]$ where $\pm V_m$ are the biasing voltages of the operational amplifiers used to implement the activation function of the neuron.

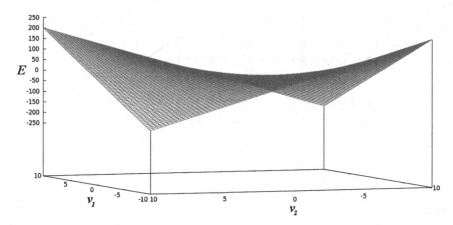

Fig. 3.2 Typical energy function plot for the Hopfield network for a 2 neuron system for the case of symmetric weight matrix **W** with zero diagonal elements

Therefore, the standard Hopfield network is not suited for the task of solving linear equations because of the following reasons:

- Equation (3.2) demands that the weight matrix must have zero diagonal elements
- Equation (3.3) demands that the weights matrix must be symmetric
- The minima of the energy function lie at the corners of the hypercube and cannot be made to occur at a point of interest, for example, the solution point of the system of linear equations

Hence, it is clear that suitable modifications need to be incorporated in the standard Hopfield network to make it amenable for the task of solving systems of simultaneous linear equations. For instance, interchanging the inverting and non-inverting inputs of the operational amplifiers in the standard Hopfield network would cause the energy function of (3.4) to become

$$E = \frac{1}{2} \sum_i \sum_j W_{ij} v_i v_j + \sum_i i_i v_i - \sum_i \frac{1}{R_i} \int_0^{v_i} g_i^{-1}(\mathbf{v}) \, d\mathbf{v} \qquad (3.5)$$

The above modification coupled with allowance of self-feedback in the neurons i.e, relaxation in the condition of (3.2) causes the minimum of the energy function to occur at the center of the hypercube. For a 2–neuron system, the typical energy function plot is presented in Fig. 3.3.

Although the suitably modified Hopfield network, as discussed above, can be made to attain a unique minimum in the energy function, the minimum exists at the center of the hypercube i.e, at $v_i = 0$, for all i. Therefore, the network is still not suitable for application in the task of solution of linear equations since for a network to be able to solve a system of linear equations, the minimum in the energy function must correspond the the solution point of the system of linear equations, which may

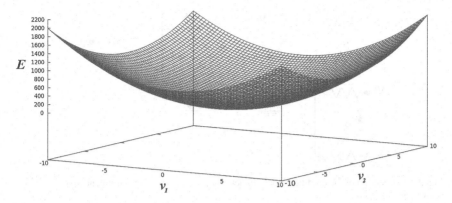

Fig. 3.3 Typical energy function plot for the Hopfield network for a 2 neuron system

not necessarily be at the origin. The next section presents further modifications to make the network suitable for solving linear equations.

3.2.2 Modified Hopfield Network for Solving Linear Equations

This section deals with the description of a modified Hopfield neural network that is suitable for the solution of a system of linear equations. Let the simultaneous linear equations to be solved be

$$\mathbf{AV} = \mathbf{B} \tag{3.6}$$

where

$$\mathbf{A} = \begin{bmatrix} a_{11} & a_{12} & \dots & a_{1n} \\ a_{21} & a_{22} & \dots & a_{2n} \\ \vdots & \vdots & \dots & \vdots \\ a_{n1} & a_{n2} & \dots & a_{nn} \end{bmatrix} \tag{3.7}$$

$$\mathbf{B} = \begin{bmatrix} b_1 \\ b_2 \\ \vdots \\ b_n \end{bmatrix} \tag{3.8}$$

$$\mathbf{V} = \begin{bmatrix} V_1 \\ V_2 \\ \vdots \\ V_n \end{bmatrix} \tag{3.9}$$

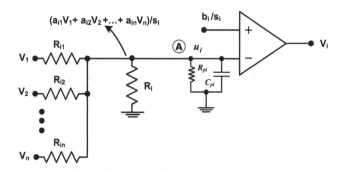

Fig. 3.4 ith neuron of the modified Hopfield neural network for solving simultaneous linear equations in n-variables

where $V_1, V_2, \ldots, V_n$ are the variables and a_{ij} and b_i are constants. Since a voltage-mode linear equation solver is presented, the decision variables are designated as voltages $V_1, V_2, \ldots, V_n$ to correspond to the output states of the neurons. It will be assumed that the coefficient matrix **A** is invertible, and hence, the system of linear equations (3.6) is consistent and not under-determined. In other words, the linear system (3.6) has a uniquely determined solution. Moreover, the solution should lie within the operating region of the neural circuit i.e, inside the hypercube defined by $|V_i| \leq V_m$ ($i = 1, 2, \ldots, n$).

The ith neuron of the voltage-mode modified Hopfield neural network based circuit for solving the system of linear equations (3.6) is presented in Fig. 3.4. It is to be noted that the output of the neuron is now designated as V_i (instead of v_i in Fig. 3.1) to conform to the notation used for decision variables in (3.9). R_{pi} and C_{pi} are the parasitic resistance and capacitance of the opamp corresponding to the ith neuron. These parasitic components are included to model the dynamic nature of the opamp. As can be seen from Fig. 3.4, individual equations from the set of equations (3.6) are scaled by a factor s_i before application to the neuron amplifiers. This scaling is done to ensure that all $[a_{ij}/s_i]$ coefficients are less than unity thereby facilitating their implementation by passive voltage dividers. As is explained later in this section, the scaling factors may be chosen independently for all the equations. In that case, the scaling factor for the ith equation, s_i, would be such that

$$s_i \geq \sum_{j=1}^{n} a_{ij} \tag{3.10}$$

All equations may also be scaled by the same factor s, which must then be the greatest of all scaling factors chosen for individual equations i.e,

$$s = \max(s_i); \text{ for all } i \tag{3.11}$$

Node equation for node 'A' gives the equation of motion of the i th neuron as

$$C_{pi}\frac{du_i}{dt} = \frac{V_1}{R_{i1}} + \frac{V_2}{R_{i2}} + \cdots + \frac{V_n}{R_{in}} - u_i\left[\frac{1}{R_{eqv,i}}\right] \tag{3.12}$$

where u_i is the internal state of the i th neuron, and

$$R_{eqv,i} = R_{i1}\,\|R_{i2}\|\,R_{in},\,\ldots,\,\|R_i\|\,R_{pi} \tag{3.13}$$

In order to arrive at a valid energy function for the modified Hopfield neural network for solving linear equations presented in Fig. 3.4, we proceed as follows. The relationship between u_i and V_i for the operational amplifier in Fig. 3.4 is given by

$$V_i = g_i\left(\frac{b_i}{s_i} - u_i\right) \tag{3.14}$$

which can be rewritten as

$$u_i = \frac{b_i}{s_i} - g_i^{-1}(V_i) \tag{3.15}$$

Also, for such dynamical systems like the one shown in Fig. 3.4, the gradient of the energy function E is related to the time evolution of the network as given by (3.16) [12].

$$\frac{\partial E}{\partial V_i} = C_{pi}\frac{du_i}{dt}; \text{ for all } i \tag{3.16}$$

Using (3.12) and (3.16), along with (3.3), the energy function corresponding to the network of Fig. 3.4 can be written as

$$E = \frac{1}{2}\sum_i\sum_j W_{ij}V_iV_j - \sum_i\frac{b_i/s_i}{R_{eqv,i}}V_i + \sum_i\frac{1}{R_{eqv,i}}\int_0^{V_i} g_i^{-1}(\mathbf{V})\,d\mathbf{V} \tag{3.17}$$

The last term in (3.17) is negligible in comparison to the first two terms and is usually neglected [11, 12]. Therefore, the energy function expression can be simplified to

$$E = \frac{1}{2}\sum_i\sum_j W_{ij}V_iV_j - \sum_i\frac{b_i/s_i}{R_{eqv,i}}V_i \tag{3.18}$$

From (3.18), it is evident that the minimum of the energy function will also be governed by the values of the elements of the vector $\mathbf{B}$ in (3.8), and therefore, the minimum will now not be at the center of the hypercube as was the case with the energy function in (3.5).

The stationary point of the energy function of (3.18) can be found by setting

$$\frac{\partial E}{\partial V_i} = 0; \quad i = 1, 2, \ldots, n \tag{3.19}$$

from where, we get

$$\frac{V_1}{R_{i1}} + \frac{V_2}{R_{i2}} + \cdots + \frac{V_n}{R_{in}} - \frac{b_i/s_i}{R_{eqv,i}} = 0; \quad i = 1, 2, \ldots, n \tag{3.20}$$

The system of linear equations of (3.6) can be written as

$$\frac{a_{i1}V_1}{s_i} + \frac{a_{i2}V_2}{s_i} + \cdots + \frac{a_{in}V_n}{s_i} - \frac{b_i}{s_i} = 0; \quad i = 1, 2, \ldots, n \tag{3.21}$$

In order for the stationary point of the energy function of (3.18), as found in (3.20) to coincide with the solution point of the system of linear equations given in (3.21), the values of the resistances in the network should be

$$R_{ij} = \frac{s_i}{a_{ij}}; \quad \begin{matrix} i = 1, 2, \ldots, n \\ j = 1, 2, \ldots, n \end{matrix} \tag{3.22}$$

The value of the resistance R_i, can be found by equating the last terms in (3.20) and (3.21), thereby yielding

$$\frac{1}{R_{eqv,i}} = 1 \tag{3.23}$$

which can be written as

$$\frac{1}{R_{i1}} + \frac{1}{R_{i2}} + \cdots + \frac{1}{R_{in}} + \frac{1}{R_i} + \frac{1}{R_{pi}} = 1 \tag{3.24}$$

Also, since R_{pi} is much larger than all the other resistances in (3.24), it can be neglected while computing the parallel equivalent of all resistances connected at the ith node, and therefore

$$\frac{1}{R_{i1}} + \frac{1}{R_{i2}} + \cdots + \frac{1}{R_{in}} + \frac{1}{R_i} = 1 \tag{3.25}$$

Substituting the values of resistances R_{ij} from (3.22) into (3.25), we get

$$R_i = \frac{s_i}{s_i - \sum_{j=1}^{n} a_{ij}} \tag{3.26}$$

From (3.26), the constraint enforced on the scaling factor as given in (3.10) can also be obtained since choosing a scaling factor in violation of (3.10) would result in a *negative* resistance according to (3.26).

Using (3.22), the values of all the weight resistances in the modified Hopfield network applied to solve linear equations can be given by

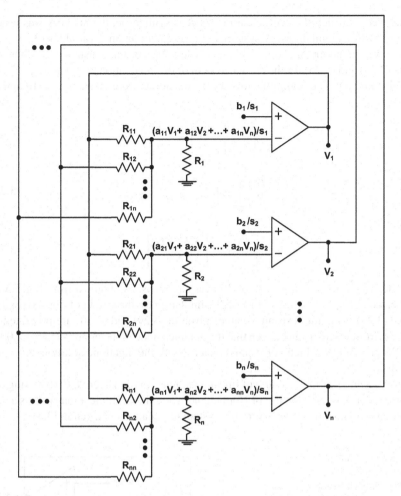

Fig. 3.5 Complete circuit of the modified Hopfield neural network applied for solving n simultaneous linear equations in n-variables

$$\begin{bmatrix} R_{11} & R_{12} & \cdots & R_{1n} \\ R_{21} & R_{22} & \cdots & R_{2n} \\ \vdots & \vdots & \ddots & \vdots \\ R_{n1} & R_{n2} & \cdots & R_{nn} \end{bmatrix} = \begin{bmatrix} \frac{1}{a_{11}/s_1} & \frac{1}{a_{12}/s_1} & \cdots & \frac{1}{a_{1n}/s_1} \\ \frac{1}{a_{21}/s_2} & \frac{1}{a_{22}/s_2} & \cdots & \frac{1}{a_{2n}/s_2} \\ \vdots & \vdots & \ddots & \vdots \\ \frac{1}{a_{n1}/s_n} & \frac{1}{a_{n2}/s_n} & \cdots & \frac{1}{a_{nn}/s_n} \end{bmatrix} \qquad (3.27)$$

The complete symmetric Hopfield neural network for solving a system of n simultaneous linear equations in n variables obtained by employing neurons as presented in Fig. 3.4 is shown is Fig. 3.5.

In order to ascertain that the modified Hopfield neural network based linear equation solver of Fig. 3.5 is able to provide solutions for systems of simultaneous linear equations for which the coefficient matrix $\mathbf{A}$ is symmetric, the network was tested using PSPICE simulations. A system of linear equations in 2 variables (3.28) was first solved using the network of Fig. 3.5. The values of the resistances as obtained from (3.27) and (3.26) for the chosen system of linear equations in 2 variables (3.28) are given below. The scaling factors for the equations were taken as $s_1 = 10$ and $s_2 = 10$.

$$\begin{bmatrix} 4 & 3 \\ 3 & 5 \end{bmatrix} \begin{bmatrix} V_1 \\ V_2 \end{bmatrix} = \begin{bmatrix} 1 \\ 9 \end{bmatrix} \tag{3.28}$$

$$\begin{bmatrix} R_{11} & R_{12} \\ R_{21} & R_{22} \end{bmatrix} = \begin{bmatrix} 2.5\,K\Omega & 3.33\,K\Omega \\ 3.33\,K\Omega & 2\,K\Omega \end{bmatrix} \tag{3.29}$$

and

$$\begin{bmatrix} R_1 \\ R_2 \end{bmatrix} = \begin{bmatrix} 3.33\,K\Omega \\ 5\,K\Omega \end{bmatrix} \tag{3.30}$$

The circuit to solve a system of 2 linear equations, as obtained from Fig. 3.5, is presented in Fig. 3.6. Results of PSPICE simulation for the circuit of Fig. 3.6 used to solve (3.28) using the resistance values given in (3.29) and (3.30) are presented in Fig. 3.7 from where it can be seen that the obtained node voltages are $V(1) = -2.00\,V$ and $V(2) = 3.00\,V$ which correspond exactly with the algebraic solution, $V_1 = -2$ and $V_2 = 3$.

Next, a three variable system of linear equations was considered. Considering the restriction imposed by the use of Hopfield network, the coefficient matrix $\mathbf{A}$ is again taken to be symmetric. The 3-variable system of equations is given in (3.31).

Fig. 3.6 The modified Hopfield neural network applied for solving simultaneous linear equations in 2 variables, with symmetric interconnection matrix

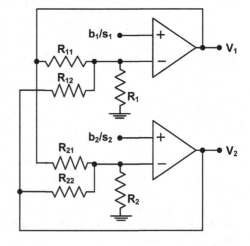

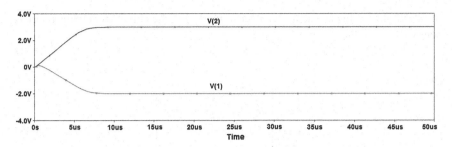

Fig. 3.7 Results of PSPICE simulation for the network of Fig. 3.6 applied to solve (3.28)

$$\begin{bmatrix} 2 & 3 & 1 \\ 3 & 5 & 2 \\ 1 & 2 & 4 \end{bmatrix} \begin{bmatrix} V_1 \\ V_2 \\ V_3 \end{bmatrix} = \begin{bmatrix} 0 \\ 1 \\ 7 \end{bmatrix} \tag{3.31}$$

PSPICE simulation for the Hopfield network based circuit used to solve (3.31) using the resistance values given in (3.32) and (3.33) yielded node voltages equal to $V(1) = -1.02$ V, $V(2) = 135\ \mu$ V and $V(3) = 1.99$ V which correspond well with the algebraic solution, $V_1 = -1$, $V_2 = 0$ and $V_3 = +2$. The scaling factors were kept as $s_1 = 6$, $s_2 = 10$ and $s_3 = 7$.

$$\begin{bmatrix} R_{11} & R_{12} & R_{13} \\ R_{21} & R_{22} & R_{23} \\ R_{31} & R_{32} & R_{33} \end{bmatrix} = \begin{bmatrix} 3\,K\Omega & 2\,K\Omega & 6\,K\Omega \\ 3.33\,K\Omega & 2\,K\Omega & 5\,K\Omega \\ 7\,K\Omega & 3.5\,K\Omega & 1.75\,K\Omega \end{bmatrix} \tag{3.32}$$

and

$$\begin{bmatrix} R_1 \\ R_2 \\ R_3 \end{bmatrix} = \begin{bmatrix} \infty \\ \infty \\ \infty \end{bmatrix} \tag{3.33}$$

The results of PSPICE simulations for both the problems considered above highlight the pitfall of Hopfield Neural Network based approach to solving linear equations. The technique works well for only those cases in which the coefficient matrix **A** is symmetric. For systems of linear equations which do not have symmetric **A**, the approach fails, and alternative solutions need to be employed.

3.3 NOSYNN-Based Neural Circuit for Solving Linear Equations

The NOSYNN-based linear equation solver discussed in this section is a viable alternative to the HNN-based solver of the previous section. The non-linear feedback approach alleviates the pitfalls associated with the Hopfield network. It shall be shown in due course that symmetricity of **A** is not an essential criterion for the

NOSYNN-based linear equation solver to work. Let the simultaneous linear equations to be solved are

$$\mathbf{AV} = \mathbf{B} \tag{3.34}$$

where

$$\mathbf{A} = \begin{bmatrix} a_{11} & a_{12} & a_{13} & \cdots & a_{1n} \\ a_{21} & a_{22} & a_{23} & \cdots & a_{2n} \\ a_{31} & a_{32} & a_{33} & \cdots & a_{3n} \\ \vdots & \vdots & \vdots & \cdots & \vdots \\ a_{n1} & a_{n2} & a_{n3} & \cdots & a_{nn} \end{bmatrix} \tag{3.35}$$

$$\mathbf{B} = \begin{bmatrix} b_1 \\ b_2 \\ b_3 \\ \vdots \\ b_n \end{bmatrix} \tag{3.36}$$

$$\mathbf{V} = \begin{bmatrix} V_1 \\ V_2 \\ V_3 \\ \vdots \\ V_n \end{bmatrix} \tag{3.37}$$

where $V_1, V_2, \ldots, V_n$ are the variables and a_{ij} and b_i are constants. It may be mentioned that (3.34) is a compact representation of (2.82) and utilizes the matrix notation. Also, since a voltage-mode linear equation solver is presented, the decision variables are now designated as voltages $V_1, V_2, \ldots, V_n$ to correspond to the output states of the neurons. It will be assumed that the coefficient matrix $\mathbf{A}$ is invertible, and hence, the system of linear equations (3.34) is consistent and not under-determined. In other words, the linear system (3.34) has a uniquely determined solution.

The NOSYNN-based neural-network based circuit to solve the system of equations of (3.34) is presented in Fig. 3.8, from where it can be seen that individual equations from the set of equations to be solved are passed through non-linear synapses which are realized using voltage-mode comparators. The outputs of the comparators are fed to neurons having weighted inputs. These weighted neurons are realized by using opamps where the resistors R_{ij} act as weights. R_{pi} and C_{pi} are the input resistance and capacitance of the opamp corresponding to the ith neuron. These parasitic components are included to model the dynamic nature of the opamp.

In an idealized scenario, the voltage-mode comparators would be different from the neuronal amplifier, which is taken as an operational amplifier. In that case, the delay of the comparator would be much smaller than the delay of the neuronal amplifier. In fact, all subsequent discussions regarding neuronal dynamics presented in this text assume this fact *viz.* all the comparators do not possess any delay and

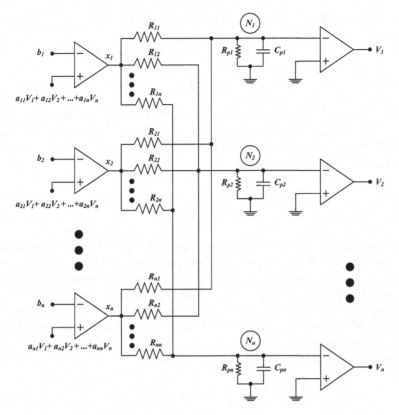

Fig. 3.8 ith neuron of the NOSYNN-based feedback neural network circuit to solve simultaneous linear equations in n-variables

all the delay of the neuron is modelled at the neuron amplifier. It is this non-linear nature of feedback which makes it different from the Hopfield neural Network and its variants (which have linear feedback, typically implemented using resistors). This also results in an energy function which is significantly different from the quadratic form of the energy function associated with HNN-based circuits.

Solution of two simultaneous linear equations in two variables
In order to understand the operation of the circuit, and also to obtain the energy function of the network, it is advisable to start from a simpler version of the general network shown in Fig. 3.8. For the sake of simplicity, we shall start with a circuit capable of solving two simultaneous linear equations in two variables. From Fig. 3.8, it is readily apparent that a circuit to solve a two variable problem would require 2 neuron amplifiers and 2 comparators. The inputs of the first comparator would be: a voltage equal to b_1 at the inverting input terminal and $(a_{11}V_1 + a_{12}V_2)$ at the non-inverting input. The output of this first comparator is x_1. Similarly, for the second voltage-mode comparator, the inputs and outputs would be b_2 at the inverting input

terminal and $(a_{21}V_1 + a_{22}V_2)$ at the non-inverting input terminal. The output of the second comparator will be x_2.

The output of the ith comparator in Fig. 3.8 can be modelled as

$$x_i = V_m tanh\beta (a_{i1}V_1 + a_{i2}V_2 + \cdots + a_{in}V_n - b_i) \qquad (3.38)$$

where β is the open-loop gain of the comparator (practically very high), $\pm Vm$ are the output voltage levels of the comparator and $V_1, V_2, \ldots, V_n$ are the neuron outputs. Therefore, for a two variable linear equation solver, the outputs of the two comparators would be given by

$$x_1 = V_m \tanh \beta(a_{11}V_1 + a_{12}V_2 - b_1) \qquad (3.39)$$
$$x_2 = V_m \tanh \beta(a_{21}V_1 + a_{22}V_2 - b_2) \qquad (3.40)$$

The resistances connected at the inverting terminal of the opamp emulating the first neuron will be R_{11} (from the output of the first comparator) and R_{21} from the output of the second comparator. Similarly, resistances R_{12} (from the output of the first comparator) and R_{22} (from the output of the second neuron) are connected to the inverting terminal of the opamp serving as an electronic model of the second neuron. The dynamics of this two neuron circuit can be understood by writing the equations of motion of the two neurons in state space.

At the input of the first neuron amplifier, we have

$$\frac{x_1 - u_1}{R_{11}} + \frac{x_2 - u_1}{R_{21}} = C_{p1}\frac{du_1}{dt} + \frac{u_1}{R_{p1}} \qquad (3.41)$$

which can be rearranged as

$$C_{p1}\frac{du_1}{dt} = \frac{x_1}{R_{11}} + \frac{x_2}{R_{21}} - u_1\left[\frac{1}{R_{11}} + \frac{1}{R_{21}} + \frac{1}{R_{p1}}\right] \qquad (3.42)$$

and can be simplified to

$$C_{p1}\frac{du_1}{dt} = \frac{x_1}{R_{11}} + \frac{x_2}{R_{21}} - \frac{u_1}{R_{eff1}} \qquad (3.43)$$

where

$$\frac{1}{R_{eff1}} = \frac{1}{R_{11}} + \frac{1}{R_{21}} + \frac{1}{R_{p1}} \qquad (3.44)$$

Substituting the value of x_1 and x_2 from (3.39) and (3.40) respectively in (3.43), we get

$$C_{p1}\frac{du_1}{dt} = \frac{V_m \tanh \beta(a_{11}V_1 + a_{12}V_2 - b_1)}{R_{11}}$$
$$+ \frac{V_m \tanh \beta(a_{21}V_1 + a_{22}V_2 - b_2)}{R_{21}} - \frac{u_1}{R_{eff1}} \tag{3.45}$$

A similar analysis for the second neuron yields the following

$$C_{p2}\frac{du_2}{dt} = \frac{V_m \tanh \beta(a_{11}V_1 + a_{12}V_2 - b_2)}{R_{12}}$$
$$+ \frac{V_m \tanh \beta(a_{21}V_1 + a_{22}V_2 - b_2)}{R_{22}} - \frac{u_2}{R_{eff2}} \tag{3.46}$$

where

$$\frac{1}{R_{eff2}} = \frac{1}{R_{12}} + \frac{1}{R_{22}} + \frac{1}{R_{p2}} \tag{3.47}$$

Furthermore, dynamical systems such as the one discussed here have an energy function which is related to the equation of motion of the two neurons as

$$\frac{\partial E}{\partial V_1} = KC_{p1}\frac{du_1}{dt}; \quad \frac{\partial E}{\partial V_2} = KC_{p2}\frac{du_2}{dt} \tag{3.48}$$

where K is a multiplicative constant of proportionality with the dimensions of resistance [13, 14]. From (3.45), (3.46) and (3.48), an expression of the energy function may be obtained in one of the following two ways:

- Mathematical approach
- Intelligent guesswork

Following the second approach, one candidate for the energy function could be

$$E_{2var} = \frac{V_m}{\beta} \ln \cosh \beta(a_{11}V_1 + a_{12}V_2 - b_1) + \frac{V_m}{\beta} \ln \cosh \beta(a_{21}V_1 + a_{22}V_2 - b_2)$$
$$- \frac{1}{R_{eff1}} \int_0^{V_1} u_1 dV_1 - \frac{1}{R_{eff2}} \int_0^{V_2} u_2 dV_2 \tag{3.49}$$

The last two terms in (3.49) are only significant near the saturating values of the opamp and can otherwise be neglected for all operational voltages below the saturation voltage of the opamp [13, 14]. This results in a somewhat simpler expression of the energy function, as given in (3.50)

$$E_{2var,simplified} = \frac{V_m}{\beta} \ln \cosh \beta(a_{11}V_1 + a_{12}V_2 - b_1)$$
$$+ \frac{V_m}{\beta} \ln \cosh \beta(a_{21}V_1 + a_{22}V_2 - b_2) \tag{3.50}$$

To verify whether the chosen energy function can indeed be associated with the two variable linear equation solver, we find $\partial E/\partial V_1$ and $\partial E/\partial V_2$ for the expression in (3.50) and then compare the result with the right-hand sides of (3.48), as shown below.

$$
\begin{aligned}
\frac{\partial E}{\partial V_1} &= a_{11} V_m \tanh \beta(a_{11} V_1 + a_{12} V_2 - b_1) + a_{21} V_m \tanh \beta(a_{21} V_1 + a_{22} V_2 - b_2) \\
&= K C_{p1} \frac{du_1}{dt} = K \frac{V_m \tanh \beta(a_{11} V_1 + a_{12} V_2 - b_1)}{R_{11}} \\
&\quad + K \frac{V_m \tanh \beta(a_{21} V_1 + a_{22} V_2 - b_2)}{R_{21}}
\end{aligned}
\tag{3.51}
$$

From (3.51), it can be observed that

$$
R_{11} = \frac{K}{a_{11}}; \quad R_{21} = \frac{K}{a_{21}}
$$

Therefore, the constant of proportionality and the coefficients of the first variable V_1 in the set of linear equations to be solved, serve to set the actual values of the resistance connected at the input of the first neuron. A suitable value of K may be chosen and then the values of R_{11} and R_{21} can be found easily.

A similar analysis is needed for the second neuron, wherein we start by finding $\partial E/\partial V_2$ and then equating it with the right-hand side of the second equality in (3.48).

$$
\begin{aligned}
\frac{\partial E}{\partial V_2} &= a_{12} V_m \tanh \beta(a_{11} V_1 + a_{12} V_2 - b_1) + a_{22} V_m \tanh \beta(a_{21} V_1 + a_{22} V_2 - b_2) \\
&= K C_{p2} \frac{du_2}{dt} = K \frac{V_m \tanh \beta(a_{11} V_1 + a_{12} V_2 - b_1)}{R_{12}} \\
&\quad + K \frac{V_m \tanh \beta(a_{21} V_1 + a_{22} V_2 - b_2)}{R_{22}}
\end{aligned}
\tag{3.52}
$$

From (3.52), it can be observed that

$$
R_{12} = \frac{K}{a_{12}}; \quad R_{22} = \frac{K}{a_{22}}
$$

Therefore, for the second neuron too, the constant of proportionality and the coefficients of the second variable V_2 in the set of linear equations to be solved, serve to set the actual values of the resistance connected at the input of the second neuron. A suitable value of K may be chosen and then the values of R_{12} and R_{22} can be found easily.

As an example, consider the following set of linear equations in 2 variables

$$
\begin{bmatrix} 2 & 1 \\ 3 & 2 \end{bmatrix} \begin{bmatrix} V_1 \\ V_2 \end{bmatrix} = \begin{bmatrix} 5 \\ 10 \end{bmatrix}
\tag{3.53}
$$

In order to find the values of resistances that need to be used, we first choose $K = 6\,K\Omega$, and then compute the values of the synaptic resistances as

$$R_{11} = \frac{K}{a_{11}} = \frac{6\,K\Omega}{2} = 3\,K\Omega$$

$$R_{21} = \frac{K}{a_{21}} = \frac{6\,K\Omega}{3} = 2\,K\Omega$$

$$R_{12} = \frac{K}{a_{12}} = \frac{6\,K\Omega}{1} = 6\,K\Omega$$

$$R_{22} = \frac{K}{a_{22}} = \frac{6\,K\Omega}{2} = 3\,K\Omega$$

Another point that is worth noting is that some additional resistances would be required to generate the voltages to be applied at the non-inverting inputs of the two comparators *viz.* $a_{11}V_1 + a_{12}V_2$ and $a_{21}V_1 + a_{22}V_2$. The requisite voltage may be obtained by simple weighted summers. However, the use of active circuits for voltage summation, such as that required here, would result in significant increase in the overall circuit complexity. To keep the circuit complexity low, the voltage summation is achieved by using passive voltage dividers. The technique is explained in detail later in this chapter.

Solving three simultaneous linear equations in three variables

Next, a neural network capable of solving three linear equations in three variable is discussed. As was done for the previous case, the circuit obtained from the generalized circuit presented in Fig. 3.8 is considered. The resultant circuit, shown in Fig. 3.9 contains three neuron amplifiers (realized using opamps) and three comparators (implemented using opamps) along with nine synaptic resistances.

To understand the operation of the circuit of Fig. 3.9, we shall proceed on similar lines as was done for the case of a two variable linear equation solver. The outputs of the three comparators are given by

$$x_1 = V_m \tanh \beta(a_{11}V_1 + a_{12}V_2 + a_{13}V_3 - b_1) \tag{3.54}$$

$$x_2 = V_m \tanh \beta(a_{21}V_1 + a_{22}V_2 + a_{23}V_3 - b_2) \tag{3.55}$$

$$x_3 = V_m \tanh \beta(a_{31}V_1 + a_{32}V_2 + a_{33}V_3 - b_3) \tag{3.56}$$

These voltages at the outputs of the comparators cause currents to flow in the synaptic resistance connected between the output of each comparator and the input of each neuronal amplifier. The currents arriving at the input of each neuronal amplifier from the various synapses are then added and this combined current is responsible for deciding the input (u_i) as well as the output (V_i) state of the particular neuron. At the input of the first neuron amplifier, we have

$$\frac{x_1 - u_1}{R_{11}} + \frac{x_2 - u_1}{R_{21}} + \frac{x_3 - u_1}{R_{31}} = C_{p1}\frac{du_1}{dt} + \frac{u_1}{R_{p1}} \tag{3.57}$$

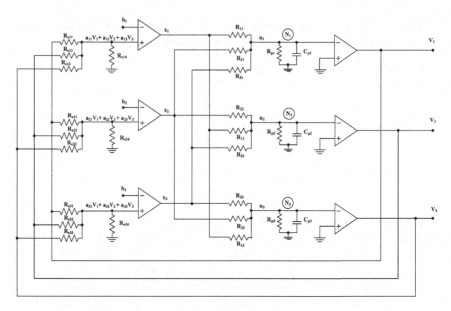

Fig. 3.9 The NOSYNN-based circuit applied to a 3—variable problem

which can be rearranged as

$$C_{p1}\frac{du_1}{dt} = \frac{x_1}{R_{11}} + \frac{x_2}{R_{21}}\frac{x_3}{R_{31}} - u_1\left[\frac{1}{R_{11}} + \frac{1}{R_{21}} + \frac{1}{R_{31}} + \frac{1}{R_{p1}}\right] \qquad (3.58)$$

and can be simplified to

$$C_{p1}\frac{du_1}{dt} = \frac{x_1}{R_{11}} + \frac{x_2}{R_{21}} + \frac{x_3}{R_{31}} - \frac{u_1}{R_{eff1}} \qquad (3.59)$$

where

$$\frac{1}{R_{eff1}} = \frac{1}{R_{11}} + \frac{1}{R_{21}} + + \frac{1}{R_{31}}\frac{1}{R_{p1}} \qquad (3.60)$$

Substituting the value of x_1, x_2 and x_3 from (3.54), (3.55) and (3.56), in (3.59), we get

$$C_{p1}\frac{du_1}{dt} = \frac{V_m \tanh \beta(a_{11}V_1 + a_{12}V_2 + a_{13}V_3 - b_1)}{R_{11}}$$
$$+ \frac{V_m \tanh \beta(a_{21}V_1 + a_{22}V_2 + a_{23}V_3 - b_2)}{R_{21}}$$
$$+ \frac{V_m \tanh \beta(a_{31}V_1 + a_{32}V_2 + a_{33}V_3 - b_3)}{R_{31}} - \frac{u_1}{R_{eff1}} \qquad (3.61)$$

A similar analysis for the second neuron yields the following

$$C_{p2}\frac{du_2}{dt} = \frac{V_m \tanh \beta(a_{11}V_1 + a_{12}V_2 + a_{13}V_3 - b_1)}{R_{12}}$$

$$+ \frac{V_m \tanh \beta(a_{21}V_1 + a_{22}V_2 + a_{23}V_3 - b_2)}{R_{22}}$$

$$+ \frac{V_m \tanh \beta(a_{31}V_1 + a_{32}V_2 + a_{33}V_3 - b_3)}{R_{32}} - \frac{u_2}{R_{eff2}} \quad (3.62)$$

where

$$\frac{1}{R_{eff2}} = \frac{1}{R_{12}} + \frac{1}{R_{22}} + \frac{1}{R_{32}} + \frac{1}{R_{p2}} \quad (3.63)$$

And finally for the third neuron, we have

$$C_{p3}\frac{du_3}{dt} = \frac{V_m \tanh \beta(a_{11}V_1 + a_{12}V_2 + a_{13}V_3 - b_1)}{R_{13}}$$

$$+ \frac{V_m \tanh \beta(a_{21}V_1 + a_{22}V_2 + a_{23}V_3 - b_2)}{R_{23}}$$

$$+ \frac{V_m \tanh \beta(a_{31}V_1 + a_{32}V_2 + a_{33}V_3 - b_3)}{R_{33}} - \frac{u_3}{R_{eff3}} \quad (3.64)$$

where

$$\frac{1}{R_{eff3}} = \frac{1}{R_{13}} + \frac{1}{R_{23}} + \frac{1}{R_{33}} + \frac{1}{R_{p3}} \quad (3.65)$$

Furthermore, dynamical systems such as the one discussed here have an energy function which is related to the equation of motion of the three neurons as

$$\frac{\partial E}{\partial V_1} = KC_{p1}\frac{du_1}{dt}$$

$$\frac{\partial E}{\partial V_2} = KC_{p2}\frac{du_2}{dt}$$

$$\frac{\partial E}{\partial V_3} = KC_{p3}\frac{du_3}{dt} \quad (3.66)$$

where K is a multiplicative constant of proportionality with the dimensions of resistance [13, 14].

To obtain the energy function associated with the three neuron network, we shall proceed by the 'intelligent guessing' technique which was adopted with success in the previous case. One possible energy function, E, the partial derivatives of which, with respect to the decision variables (V_1, V_2 and V_3), correspond with the neuronal states given in (3.61) through (3.64), could be

$$E_{3var} = \frac{V_m}{\beta} \ln \cosh \beta(a_{11}V_1 + a_{12}V_2 + a_{13}V_3 - b_1)$$

$$+ \frac{V_m}{\beta} \ln \cosh \beta(a_{21}V_1 + a_{22}V_2 + a_{23}V_3 - b_2)$$

$$+ \frac{V_m}{\beta} \ln \cosh \beta(a_{31}V_1 + a_{32}V_2 + a_{33}V_3 - b_3)$$

$$- \frac{1}{R_{eff1}} \int_0^{V_1} u_1 dV_1 - \frac{1}{R_{eff2}} \int_0^{V_2} u_2 dV_2 - \frac{1}{R_{eff3}} \int_0^{V_3} u_3 dV_3 \quad (3.67)$$

The last two terms in (3.67) are only significant near the saturating values of the opamp and can be neglected for all operational voltages below the saturation voltage of the opamp [13, 14]. This results in a somewhat simpler expression of the energy function, as given in (3.68)

$$E_{3var,simplified} = \frac{V_m}{\beta} \ln \cosh \beta(a_{11}V_1 + a_{12}V_2 + a_{13}V_3 - b_1)$$

$$+ \frac{V_m}{\beta} \ln \cosh \beta(a_{21}V_1 + a_{22}V_2 + a_{23}V_3 - b_2)$$

$$+ \frac{V_m}{\beta} \ln \cosh \beta(a_{31}V_1 + a_{32}V_2 + a_{33}V_3 - b_3) \quad (3.68)$$

To verify whether our chosen energy function can indeed be associated with the three variable linear equation solver, we find $\partial E/\partial V_1$, $\partial E/\partial V_2$ and $\partial E/\partial V_3$ for the expression in (3.68) and then compare the result with the right-hand sides of (3.66), as shown below.

$$\frac{\partial E}{\partial V_1} = a_{11}V_m \tanh \beta(a_{11}V_1 + a_{12}V_2 + a_{13}V_3 - b_1)$$

$$+ a_{21}V_m \tanh \beta(a_{21}V_1 + a_{22}V_2 + a_{23}V_3 - b_2)$$

$$+ a_{31}V_m \tanh \beta(a_{31}V_1 + a_{32}V_2 + a_{33}V_3 - b_3)$$

$$= KC_{p1}\frac{du_1}{dt} = K\frac{V_m \tanh \beta(a_{11}V_1 + a_{12}V_2 + a_{13}V_3 - b_1)}{R_{11}}$$

$$+ K\frac{V_m \tanh \beta(a_{21}V_1 + a_{22}V_2 + a_{23}V_3 - b_2)}{R_{21}}$$

$$+ K\frac{V_m \tanh \beta(a_{31}V_1 + a_{32}V_2 + a_{33}V_3 - b_3)}{R_{31}} \quad (3.69)$$

From (3.69), it can be observed that

$$R_{11} = \frac{K}{a_{11}}$$

$$R_{21} = \frac{K}{a_{21}}$$

$$R_{31} = \frac{K}{a_{31}}$$

Therefore, the constant of proportionality and the coefficients of the first variable V_1 in the set of linear equations to be solved, serve to set the actual values of the resistance connected at the input of the first neuron. A suitable value of K may be chosen and then the values of R_{11}, R_{21} and R_{31} can be found easily.

A similar analysis is needed for the second neuron, wherein we start by finding $\partial E/\partial V_2$ and then equating it with the right-hand side of the second equality in (3.66).

$$
\begin{aligned}
\frac{\partial E}{\partial V_2} &= a_{12}V_m \tanh \beta(a_{11}V_1 + a_{12}V_2 + a_{13}V_3 - b_1) \\
&\quad + a_{22}V_m \tanh \beta(a_{21}V_1 + a_{22}V_2 + a_{23}V_3 - b_2) \\
&\quad + a_{32}V_m \tanh \beta(a_{31}V_1 + a_{32}V_2 + a_{33}V_3 - b_3) \\
&= KC_{p2}\frac{du_2}{dt} = K\frac{V_m \tanh \beta(a_{11}V_1 + a_{12}V_2 + a_{13}V_3 - b_1)}{R_{12}} \\
&\quad + K\frac{V_m \tanh \beta(a_{21}V_1 + a_{22}V_2 + a_{23}V_3 - b_2)}{R_{22}} \\
&\quad + K\frac{V_m \tanh \beta(a_{31}V_1 + a_{32}V_2 + a_{33}V_3 - b_3)}{R_{32}}
\end{aligned}
\tag{3.70}
$$

From (3.70), it can be observed that

$$R_{12} = \frac{K}{a_{12}}$$

$$R_{22} = \frac{K}{a_{22}}$$

$$R_{32} = \frac{K}{a_{32}}$$

Therefore, for the second neuron too, the constant of proportionality and the coefficients of the second variable V_2 in the set of linear equations to be solved, serve to set the actual values of the resistance connected at its input. A suitable value of K may be chosen and then the values of R_{12}, R_{22} and R_{32} can be found readily.

Lastly, for the second neuron, we start by finding $\partial E/\partial V_3$ and then equating it with the right-hand side of the third equality in (3.66).

$$\frac{\partial E}{\partial V_3} = a_{13} V_m \tanh \beta(a_{11} V_1 + a_{12} V_2 + a_{13} V_3 - b_1)$$

$$+ a_{23} V_m \tanh \beta(a_{21} V_1 + a_{22} V_2 + a_{23} V_3 - b_2)$$
$$+ a_{33} V_m \tanh \beta(a_{31} V_1 + a_{32} V_2 + a_{33} V_3 - b_3)$$
$$= K C_{p3} \frac{du_3}{dt} = K \frac{V_m \tanh \beta(a_{11} V_1 + a_{12} V_2 + a_{13} V_3 - b_1)}{R_{13}}$$
$$+ K \frac{V_m \tanh \beta(a_{21} V_1 + a_{22} V_2 + a_{23} V_3 - b_2)}{R_{23}}$$
$$+ K \frac{V_m \tanh \beta(a_{31} V_1 + a_{32} V_2 + a_{33} V_3 - b_3)}{R_{33}} \quad\quad (3.71)$$

From (3.71), it can be observed that

$$R_{13} = \frac{K}{a_{13}}$$
$$R_{23} = \frac{K}{a_{23}}$$
$$R_{33} = \frac{K}{a_{33}}$$

Therefore, for the third neuron too, the constant of proportionality and the coefficients of the third variable V_3 in the set of linear equations to be solved, serve to set the actual values of the resistance connected at its input. A suitable value of K may be chosen and then the values of R_{13}, R_{23} and R_{33} can be found easily.

As an example, consider the following set of linear equations in 3 variables

$$\begin{bmatrix} 2 & 3 & 1 \\ 1 & 2 & 3 \\ 4 & 1 & 2 \end{bmatrix} \begin{bmatrix} V_1 \\ V_2 \\ V_3 \end{bmatrix} = \begin{bmatrix} 2 \\ -1 \\ -1 \end{bmatrix} \quad\quad (3.72)$$

In order to find the values of resistances that need to be used, we first choose $K = 12\ K\Omega$, and then compute the values of the synaptic resistances as

$$R_{11} = \frac{K}{a_{11}} = \frac{12\ K\Omega}{2} = 6\ K\Omega$$
$$R_{21} = \frac{K}{a_{21}} = \frac{12\ K\Omega}{1} = 12\ K\Omega$$
$$R_{31} = \frac{K}{a_{31}} = \frac{12\ K\Omega}{4} = 3\ K\Omega$$
$$R_{12} = \frac{K}{a_{12}} = \frac{12\ K\Omega}{3} = 2\ K\Omega$$
$$R_{22} = \frac{K}{a_{22}} = \frac{12\ K\Omega}{2} = 6\ K\Omega$$

$$R_{32} = \frac{K}{a_{32}} = \frac{12\,K\Omega}{1} = 12\,K\Omega$$

$$R_{13} = \frac{K}{a_{13}} = \frac{12\,K\Omega}{1} = 12\,K\Omega$$

$$R_{23} = \frac{K}{a_{23}} = \frac{12\,K\Omega}{3} = 4\,K\Omega$$

$$R_{33} = \frac{K}{a_{33}} = \frac{12\,K\Omega}{2} = 6\,K\Omega$$

It should be obvious to the reader that a prudent selection of K helps to obtain reasonable values of the resistors. On the other hand, a randomly selected value of K may result in inappropriate values of resistors. For instance, if in the three variable example, K is kept as $1\,K\Omega$, then the individual resistance values would be

$$R_{11} = \frac{K}{a_{11}} = \frac{1\,K\Omega}{2} = 500\,\Omega$$

$$R_{21} = \frac{K}{a_{21}} = \frac{1\,K\Omega}{1} = 1\,K\Omega$$

$$R_{31} = \frac{K}{a_{31}} = \frac{1\,K\Omega}{4} = 250\,\Omega$$

$$R_{12} = \frac{K}{a_{12}} = \frac{1\,K\Omega}{3} = 333.33\,\Omega$$

$$R_{22} = \frac{K}{a_{22}} = \frac{1\,K\Omega}{2} = 500\,\Omega$$

$$R_{32} = \frac{K}{a_{32}} = \frac{1\,K\Omega}{1} = 1\,K\Omega$$

$$R_{13} = \frac{K}{a_{13}} = \frac{1\,K\Omega}{1} = 1\,K\Omega$$

$$R_{23} = \frac{K}{a_{23}} = \frac{1\,K\Omega}{3} = 333.33\,\Omega$$

$$R_{33} = \frac{K}{a_{33}} = \frac{1\,K\Omega}{2} = 500\,\Omega$$

The reader should be able to ascertain that the first set of resistor values (obtained by selecting $K = 12\,K\Omega$) are more amenable for a monolithic implementation as compared to the second set of resistor values (obtained for $K=1\,K\Omega$). Next, the resistive potential divider technique for obtaining input voltages for the comparators is discussed.

Generating voltages for comparator inputs

This section discusses a simple, yet effective, technique to obtain the voltages required at the non-inverting inputs of the various comparators. As mentioned before, the technique is to employ resistive voltage dividers to generate the required voltage. Consider the three variable linear equation solver of Fig. 3.9. At the input of the first comparator, four resistances are connected. The resistance R_{e11} connected the output of the first neuron V_1 to the input of the first comparator. Similarly, the resistance R_{e12} is used to connect the output of the second neuron to the input of the first comparator. The output of the third neuron is brought to the input of the first comparator via the resistance R_{e13}. Let the voltage at the non-inverting input of the first comparator be V_x, then applying KCL at that node, we get

$$\frac{V_1 - V_x}{R_{e11}} + \frac{V_2 - V_x}{R_{e12}} + \frac{V_3 - V_x}{R_{e13}} = \frac{V_x}{R_{e14}} \tag{3.73}$$

which can be rearranged as

$$\frac{V_x}{R_{e11}} + \frac{V_x}{R_{e12}} + \frac{V_x}{R_{e13}} + \frac{V_x}{R_{e14}} = \frac{V_1}{R_{e11}} + \frac{V_2}{R_{e12}} + \frac{V_3}{R_{e13}} \tag{3.74}$$

Simplifying

$$\frac{V_x}{R_1} = \frac{V_1}{R_{e11}} + \frac{V_2}{R_{e12}} + \frac{V_3}{R_{e13}} \tag{3.75}$$

where

$$\frac{1}{R_1} = \left[\frac{1}{R_{e11}} + \frac{1}{R_{e12}} + \frac{1}{R_{e13}} + \frac{1}{R_{e14}} \right] \tag{3.76}$$

Therefore,

$$V_x = \frac{R_1}{R_{e11}} V_1 + \frac{R_1}{R_{e12}} V_2 + \frac{R_1}{R_{e13}} V_3 \tag{3.77}$$

Now, for the three variable circuit, the voltage that needs to be provided at the non-inverting terminal of the first comparator is

$$V_{non-inv,1} = a_{11} V_1 + a_{12} V_2 + a_{13} V_3 \tag{3.78}$$

From (3.77) and (3.78), we have

$$\frac{R_1}{R_{e11}} V_1 + \frac{R_1}{R_{e12}} V_2 + \frac{R_1}{R_{e13}} V_3 = a_{11} V_1 + a_{12} V_2 + a_{13} V_3 \tag{3.79}$$

From (3.79), it *seems* that the values of each of the resistors connected at the input of the comparator can be found by simply equating the coefficients of V_1, V_2 and V_3 on both the sides. However, actually doing so results in

$$\frac{R_1}{R_{e11}} = a_{11} \tag{3.80}$$

$$\frac{R_1}{R_{e12}} = a_{12} \tag{3.81}$$

$$\frac{R_1}{R_{e13}} = a_{13} \tag{3.82}$$

For positive values of coefficients a_{11}, a_{11} and a_{11}, (3.80) through (3.82) result in *negative* values of the resistance R_{e14}. As an example, consider the problem of finding the values of resistances in the resistive potential divider to obtain $(4V_1 + 3V_2 + 5V_3)$ at the input of the comparator. Using (3.80) through (3.82) in combination with (3.76) results in

$$\frac{4}{R_1} + \frac{3}{R_1} + \frac{5}{R_1} + \frac{1}{R_{e14}} = \frac{1}{R_1} \tag{3.83}$$

which yields

$$\frac{1}{R_{e14}} = \frac{-11}{R_1} \tag{3.84}$$

Since negatives resistances are difficult to obtain in actual hardware (although some active circuits are available), the resistive potential divider in its present form is not suitable for our purpose. However, a simple scaling of the linear equation, with which the comparator is associated, can result in an altogether different result for an analysis similar to the one performed above. This can be explained as follows. Let the first linear equation (from the set of three) be scaled by a constant s_1. This results in

$$\frac{a_{11}}{s_1} V_1 + \frac{a_{12}}{s_1} V_2 + \frac{a_{13}}{s_1} V_3 = \frac{b_1}{s_1} \tag{3.85}$$

If (3.85) is taken as the equation to be solved, a voltage corresponding to the left-hand side of this equation needs to be provided at the non-inverting input of the first comparator. In that case, the values of the resistances can be found as

$$\frac{R_1}{R_{e11}} V_1 + \frac{R_1}{R_{e12}} V_2 + \frac{R_1}{R_{e13}} V_3 = \frac{a_{11}}{s_1} V_1 + \frac{a_{12}}{s_1} V_2 + \frac{a_{13}}{s_1} V_3 \tag{3.86}$$

Term by term comparison in (3.86) yields

$$\frac{R_1}{R_{e11}} = \frac{a_{11}}{s_1} \tag{3.87}$$

$$\frac{R_1}{R_{e12}} = \frac{a_{12}}{s_1} \tag{3.88}$$

$$\frac{R_1}{R_{e13}} = \frac{a_{13}}{s_1} \tag{3.89}$$

The reader is encouraged to verify that the problem of negative resistance value for R_{e14} will not be present in the current case. A direct consequence of the scaling of the linear equation, as done in (3.85), is that the voltage to be applied at the *inverting* terminal of the comparator also needs to be scaled down by the same factor i.e., the voltage that actually needs to be applied would be (b_1/s_1). The same holds true for the inputs of all the comparators.

A proper selection of the value of the scaling factors plays an important role in the determination of the resistance values, and therefore, on the overall working of the circuit. If for example,

$$s_1 < a_{11} + a_{12} + a_{13} \tag{3.90}$$

the problem of negative value of R_{e14} would persist. For a positive, and easily implementable, value

$$s_1 \geq a_{11} + a_{12} + a_{13} \tag{3.91}$$

The equality in (3.91) is of particular interest since in that case the resistance R_{e14} would come out to be ∞, and therefore, need not be actually realized during the fabrication of the circuit. Thus, for a minimum resistance count, all the equations should be scaled by different scaling factors, which can be found using the general expression given below.

$$s_i = \sum_{j=1}^{n} a_{ij} \tag{3.92}$$

As an example, consider the following system of three linear equations in three variables:

$$\begin{bmatrix} 2 & 3 & 4 \\ 6 & 2 & 3 \\ 4 & 1 & 2 \end{bmatrix} \begin{bmatrix} V_1 \\ V_2 \\ V_3 \end{bmatrix} = \begin{bmatrix} 20 \\ 13 \\ -15 \end{bmatrix} \tag{3.93}$$

In order to obtain the values of resistances to be connected at the inputs of the three comparators, we proceed as follows. The first equation is scaled by 9 (calculated using (3.92)), and can therefore by written as

$$\frac{2}{9}V_1 + \frac{3}{9}V_2 + \frac{4}{9}V_3 = \frac{20}{9} \tag{3.94}$$

thereby yielding the values of resistances to be connected at the input of the first comparator as

$$R_{e11} = 4.5 \, K\Omega$$
$$R_{e12} = 3 \, K\Omega$$
$$R_{e13} = 2.25 \, K\Omega$$

and the voltage to be applied at the inverting input terminal of the first comparator would be

$$b_1 = 2.22 \, V$$

The second equation should be scaled by 11, thereby yielding

$$R_{e21} = 1.83 \ K\Omega$$
$$R_{e22} = 5.5 \ K\Omega$$
$$R_{e23} = 3.66 \ K\Omega$$

and the voltage to be applied at the inverting input of the second comparator as

$$b_2 = 1.18 \ V$$

Lastly, the third equation needs to be scaled by 7, thereby resulting in

$$R_{e21} = 1.75 \ K\Omega$$
$$R_{e22} = 7 \ K\Omega$$
$$R_{e23} = 3.5 \ K\Omega$$

and the voltage to be applied at the inverting input of the second comparator as

$$b_2 = -2.14 \ V$$

It may be mentioned that instead of choosing a different scaling factor for each equation in the given system of linear equations, a more prudent option is to scale all the equations by the same factor which then must be the greatest of all the scaling factors determined for individual equations (or even more than that). In the above example, a scaling factor of max [7, 9, 11] i.e., 11 is appropriate. Or else, all equations may be scaled by a factor more than 11 (for example, 15) but that would have the following implications:

- Resistances R_{e14}, R_{e24} and R_{e34} would need to be considered
- A unnecessarily high scaling factor would translate to high values of resistances which would require larger areas during the actual integration

It should be kept in mind that the scaling of equations does not have any appreciable effect on the outputs of the comparators. This can be mathematically explained as follows. The output of the first comparator, for the case when equation-1 has not been scaled, will be

$$x_1 = V_m \tanh \beta(a_{11} V_1 + a_{12} V_2 + a_{13} V_3 - b_1) \tag{3.95}$$

and the output with scaling of equation-1 will be

$$x_1 = V_m \tanh \beta \left(\frac{a_{11}}{s_1} V_1 + \frac{a_{12}}{s_1} V_2 + \frac{a_{13}}{s_1} V_3 - \frac{b_1}{s_1} \right) \tag{3.96}$$

which can be simplified to

$$x_1 = V_m \tanh \frac{\beta}{s_1}(a_{11}V_1 + a_{12}V_2 + a_{13}V_3 - b_1) \tag{3.97}$$

For very high values of the comparator gain ($\beta \to \infty$), we have

$$V_m \tanh \frac{\beta}{s_1}(a_{11}V_1 + a_{12}V_2 + a_{13}V_3 - b_1) \approx V_m \tanh \beta(a_{11}V_1 + a_{12}V_2 + a_{13}V_3 - b_1)$$
$$\tag{3.98}$$

and therefore, the output of the comparator would be almost the same for the two cases.

Solving n linear equations in n variables

Having explained the energy function and the method to obtain the values of different resistances to be used in the circuits to solve 2 and 3 variable systems of equations, we now move on to a generic network which can be used to solve n linear equations by employing n neurons connected through n non-linear synapses. The circuit diagram for the general network has already been presented in Fig. 3.8 from where it can be seen that n operational amplifiers are required to emulate the functionality of n neurons and n voltage comparators are needed to provide the required non-linear synaptic feedback from the output of the neurons to the inputs of other neurons.

To find the values of the various resistances in Fig. 3.8, we follow a procedure similar to that employed in the case of 2 and 3 variable linear equation solvers discussed previously, and start by writing the equations of motion of the ith neuron in the state space and then associate a valid energy function with the network. Once an energy function is ascertained, the values of resistances are then determined by correlating the terms in the partial differential of the energy function with respect to each decision variable, with their counterparts in the equation of motion for each neuron.

Applying KCL at the input node for the ith neuron (N_i) in Fig. 3.8, results in the following equation of motion of the i th neuron in the state space

$$C_{pi}\frac{du_i}{dt} = \frac{x_1}{R_{1i}} + \frac{x_2}{R_{2i}} + \cdots + \frac{x_n}{R_{ni}} - u_i\left[\frac{1}{R_i}\right] \tag{3.99}$$

where u_i is the internal state of the ith neuron, and

$$R_i = R_{1i} \parallel R_{2i}, \ldots, \parallel R_{ni} \parallel R_{pi} \tag{3.100}$$

Using (3.38) in (3.99) results in

$$C_{pi}\frac{du_i}{dt} = \frac{V_m \tanh \beta(a_{11}V_1 + a_{12}V_2 + \cdots + a_{1n}V_n - b_1)}{R_{1i}}$$
$$+ \frac{V_m \tanh \beta(a_{21}V_1 + a_{22}V_2 + \cdots + a_{2n}V_n - b_2)}{R_{2i}} + \cdots$$

$$+ \frac{V_m \tanh \beta (a_{n1} V_1 + a_{n2} V_2 + \cdots + a_{nn} V_n - b_n)}{R_{ni}} - \frac{u_i}{R_i} \qquad (3.101)$$

As has been demonstrated in earlier sections, an energy function can be associated with a dynamical system whose equation of motion is known. Following an approach similar to the ones used for the case of 2 and 3 variable linear equation solver, the network in Fig. 3.8 can be associated with an Energy Function E given by

$$E = \frac{V_m}{\beta} \sum_{i=1}^{n} \ln \cosh \beta \left(\sum_{j=1}^{n} a_{ij} V_j - b_i \right) - \sum_{i=1}^{n} \frac{1}{R_i} \int_0^{V_i} u_i dV_i \qquad (3.102)$$

From (3.102), it follows that

$$\frac{\partial E}{\partial V_i} = V_m a_{1i} \tanh \beta (a_{11} V_1 + a_{12} V_2 + \cdots + a_{1n} V_n - b_1)$$

$$+ V_m a_{2i} \tanh \beta (a_{21} V_1 + a_{22} V_2 + \cdots + a_{2n} V_n - b_2) + \ldots$$

$$+ V_m a_{ni} \tanh \beta (a_{n1} V_1 + a_{n2} V_2 + \cdots + a_{nn} V_n - b_n) - u_i \left[\frac{1}{R_i} \right] \qquad (3.103)$$

Also, if E is the Energy Function, it must satisfy the following condition [13, 14].

$$\frac{\partial E}{\partial V_i} = K C_{pi} \frac{du_i}{dt} \qquad (3.104)$$

where K is a constant of proportionality and has the dimensions of resistance. Comparing (3.101) and (3.103) according to (3.104) yields

$$a_{11} = \frac{K}{R_{11}}, \quad a_{21} = \frac{K}{R_{21}}, \quad \ldots, \quad a_{n1} = \frac{K}{R_{n1}} \qquad (3.105)$$

A similar analysis yields the values of the weights for the remaining neurons as

$$\begin{bmatrix} a_{11} & a_{12} & \ldots & a_{1n} \\ a_{21} & a_{22} & \ldots & a_{2n} \\ \vdots & \vdots & \ldots & \vdots \\ a_{n1} & a_{n2} & \ldots & a_{nn} \end{bmatrix} = \begin{bmatrix} K/R_{11} & K/R_{12} & \ldots & K/R_{1n} \\ K/R_{21} & K/R_{22} & \ldots & K/R_{2n} \\ \vdots & \vdots & \ldots & \vdots \\ K/R_{n1} & K/R_{n2} & \ldots & K/R_{nn} \end{bmatrix} \qquad (3.106)$$

The values of the resistances of Fig. 3.8 can therefore be easily calculated by choosing a suitable value of K and then using (3.106).

3.3.1 Proof of the Energy Function

The time derivative of the energy function is given by

$$\frac{dE}{dt} = \sum_{i=1}^{N} \frac{\partial E}{\partial V_i} \frac{dV_i}{dt} = \sum_{i=1}^{N} \frac{\partial E}{\partial V_i} \frac{dV_i}{du_i} \frac{du_i}{dt} \qquad (3.107)$$

Using (3.104) in (3.107) we get

$$\frac{dE}{dt} = \sum_{i=1}^{N} K C_i \left(\frac{du_i}{dt} \right)^2 \frac{dV_i}{du_i} \qquad (3.108)$$

The transfer characteristics of the output opamp used in Fig. 3.8 implements the activation function of the neuron. With u_i being the internal state at the inverting terminal, it is monotonically decreasing and it can be seen that [13, 14],

$$\frac{dV_i}{du_i} \leq 0 \qquad (3.109)$$

thereby resulting in

$$\frac{dE}{dt} \leq 0 \qquad (3.110)$$

with the equality being valid for

$$\frac{du_i}{dt} = 0 \qquad (3.111)$$

Equation (3.110) shows that the energy function can never increase with time which is one of the conditions for a valid energy function. The second criterion *viz.* the energy function must have a lower bound is also satisfied for the circuit of Fig. 3.8 wherein it may be seen that $V_1, V_2, \ldots, V_n$ are all bounded (as they are the outputs of opamps) amounting to E, as given in (3.102), having a defined lower bound.

3.3.2 Stable States of the Network

Convergence of the network to the global minimum of the Energy Function, which is exactly the solution of the set of linear equations, and the fact that there are no other minima, can be shown as follows.

Simple Case: 2 linear equations in 2 variables
The simplified expression of the energy function, for the case of a 2 neuron circuit, capable of solving 2 linear equations in 2 variables, was shown to be

$$E_{2var,simplified} = \frac{V_m}{\beta} \ln \cosh \beta(a_{11}V_1 + a_{12}V_2 - b_1)$$

$$+ \frac{V_m}{\beta} \ln \cosh \beta(a_{21}V_1 + a_{22}V_2 - b_2) \tag{3.112}$$

from which it follows that

$$\frac{\partial E}{\partial V_1} = a_{11}V_m \tanh \beta(a_{11}V_1 + a_{12}V_2 - b_1)$$

$$+ a_{21}V_m \tanh \beta(a_{21}V_1 + a_{22}V_2 - b_2) \tag{3.113}$$

and

$$\frac{\partial E}{\partial V_2} = a_{12}V_m \tanh \beta(a_{11}V_1 + a_{12}V_2 - b_1)$$

$$+ a_{22}V_m \tanh \beta(a_{21}V_1 + a_{22}V_2 - b_2) \tag{3.114}$$

For a stationary point, the partial derivatives of E with respect to both the decision variables must be zero.

$$\frac{\partial E}{\partial V_1} = 0; \quad \frac{\partial E}{\partial V_2} = 0; \tag{3.115}$$

Using (3.113) and (3.114) in (3.115), we get

$$a_{11}V_m \tanh \beta(a_{11}V_1 + a_{12}V_2 - b_1) + a_{21}V_m \tanh \beta(a_{21}V_1 + a_{22}V_2 - b_2) = 0$$
$$a_{12}V_m \tanh \beta(a_{11}V_1 + a_{12}V_2 - b_1) + a_{22}V_m \tanh \beta(a_{21}V_1 + a_{22}V_2 - b_2) = 0 \tag{3.116}$$

For the sake of simplicity, we denote $V_m \tanh \beta(a_{11}V_1 + a_{12}V_2 - b_1)$ by A_1 and $V_m \tanh \beta(a_{21}V_1 + a_{22}V_2 - b_2)$ by A_2, (3.116) can be simplified to

$$a_{11}A_1 + a_{21}A_2 = 0$$
$$a_{12}A_1 + a_{22}A_2 = 0 \tag{3.117}$$

Equation (3.117) may be represented in matrix notation as

$$\begin{bmatrix} a_{11} & a_{21} \\ a_{12} & a_{22} \end{bmatrix} \begin{bmatrix} A_1 \\ A_2 \end{bmatrix} = \begin{bmatrix} 0 \\ 0 \end{bmatrix} \tag{3.118}$$

Since the above is a homogeneous system of linear equations, we can easily obtain the trivial solution which is

$$\begin{bmatrix} A_1 \\ A_2 \end{bmatrix} = \begin{bmatrix} 0 \\ 0 \end{bmatrix} \tag{3.119}$$

which can be expanded to

$$V_m \tanh \beta(a_{11} V_1 + a_{12} V_2 - b_1) = 0$$
$$V_m \tanh \beta(a_{12} V_1 + a_{22} V_2 - b_2) = 0 \qquad (3.120)$$

and then simplified to

$$a_{11} V_1 + a_{12} V_2 - b_1 = 0$$
$$a_{12} V_1 + a_{22} V_2 - b_2 = 0 \qquad (3.121)$$

It is readily verified that the stationary point obtained in (3.121) is the same as the solution of the system of equations in 2 variables. Next, a similar proof is presented for the generalized case of a NOSYNN based neural network comprising on n neurons and n synapses, employed to solve a system of n simultaneous linear equations in n variables.

General Case: n linear equations in n variables
For n variables, the second term in the energy function expression (3.102) is significant only near the saturating values of the opamp and is usually neglected [11]. The energy function can therefore be expressed as

$$E = \frac{V_m}{\beta} \sum_{i=1}^{n} \ln \cosh \beta \left(\sum_{j=1}^{n} a_{ij} V_j - b_i \right) \qquad (3.122)$$

From which it follows that

$$\frac{\partial E}{\partial V_i} = \frac{V_m}{\beta} a_{1i} \tanh \beta(a_{11} V_1 + a_{12} V_2 + \cdots + a_{1n} V_n - b_1)$$
$$+ \frac{V_m}{\beta} a_{2i} \tanh \beta(a_{21} V_1 + a_{22} V_2 + \cdots + a_{2n} V_n - b_2) + \cdots$$
$$+ \frac{V_m}{\beta} a_{ni} \tanh \beta(a_{n1} V_1 + a_{n2} V_2 + \cdots + a_{nn} V_n - b_n) \qquad (3.123)$$

For a stationary point, we have

$$\frac{\partial E}{\partial V_i} = 0 \qquad (3.124)$$

which yields,

$$a_{1i} \tanh \beta(a_{11} V_1 + a_{12} V_2 + \cdots + a_{1n} V_n - b_1)$$
$$+ a_{2i} \tanh \beta(a_{21} V_1 + a_{22} V_2 + \cdots + a_{2n} V_n - b_2) + \cdots$$
$$+ a_{ni} \tanh \beta(a_{n1} V_1 + a_{n2} V_2 + \cdots + a_{nn} V_n - b_n) = 0 \qquad (3.125)$$

Denoting

$$tanh \beta(a_{11} V_1 + a_{12} V_2 + \cdots + a_{1n} V_n - b_1) = A_1$$

$$tanh\beta(a_{21}V_1 + a_{22}V_2 + \cdots + a_{2n}V_n - b_2) = A_2$$

$$\vdots$$

$$tanh\beta(a_{n1}V_1 + a_{n2}V_2 + \cdots + a_{nn}V_n - b_n) = A_n \qquad (3.126)$$

Therefore, for a stationary point we have,

$$\begin{bmatrix} a_{11} & a_{12} & \ldots & a_{1n} \\ a_{21} & a_{22} & \ldots & a_{2n} \\ \vdots & \vdots & \ldots & \vdots \\ a_{n1} & a_{n2} & \ldots & a_{nn} \end{bmatrix} \begin{bmatrix} A_1 \\ A_2 \\ \vdots \\ A_n \end{bmatrix} = \begin{bmatrix} 0 \\ 0 \\ \vdots \\ 0 \end{bmatrix} \qquad (3.127)$$

This is a homogeneous system of linear equations in variables $A_1, A_2, ..., A_n$. Since the coefficient matrix of the set of equations (3.127) is the same as that of (3.34) which is invertible, it follows that (3.127) will have a uniquely determined solution which is the trivial solution of the homogeneous system. Therefore,

$$\begin{bmatrix} A_1 \\ A_2 \\ \vdots \\ A_n \end{bmatrix} = \begin{bmatrix} 0 \\ 0 \\ \vdots \\ 0 \end{bmatrix} \qquad (3.128)$$

which results in,

$$a_{11}V_1 + a_{12}V_2 + \cdots + a_{1n}V_n - b_1 = 0$$
$$a_{21}V_1 + a_{22}V_2 + \cdots + a_{2n}V_n - b_2 = 0$$

$$\vdots$$

$$a_{n1}V_1 + a_{n2}V_2 + \cdots + a_{nn}V_n - b_n = 0 \qquad (3.129)$$

Thus, the energy function of the NOSYNN-based neural network has a unique stationary point which coincides exactly with the solution of the given system of linear equations.

The energy function, as given in (3.122), can be visualized in 3–dimensions for a 2–variable problem and is shown in Fig. 3.10. From the plot, it can be seen that there exists only one minimum to which the network must converge. Also, since the 'walls' of the energy function are more steep than the ones encountered in the case of a quadratic energy function (in the case of a Hopfield Neural Network), the system reaches its stable state (the minimum point in Fig. 3.10) in a shorter time duration thereby providing a desirable feature of fast convergence. It needs to be reiterated that this guaranteed convergence to a unique minimum point is a very desirable feature in an energy function, and is difficult to achieve in Hopfield Neural Networks and its variants, which are plagued by the problem of convergence to ill

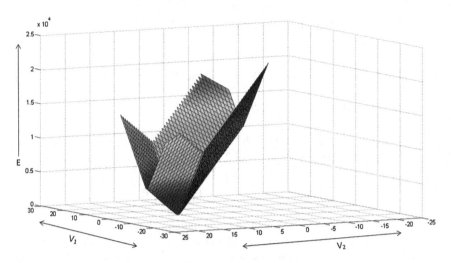

Fig. 3.10 Typical energy function plot for a 2—variable problem

defined local minima and the inability to get to a steady state (the infamous problem of oscillations).

3.3.3 Convergence of the Network

To get an estimate of the convergence time of the NOSYNN-based neural circuit for solving linear equations, we shall assume, as before, that all the delay in the neuron is modelled in the neuron. This assumes voltage comparators which are capable of providing output voltages immediately after application of inputs. For the opamps used to emulate the functionality of the neurons in Fig. 3.8, we can write

$$V_i = A(V^+ - V^-) \tag{3.130}$$

where A is the open-loop gain of the operational amplifier, and is usually very high. Since the internal state of the ith neuron in Fig. 3.8 is denoted by u_i, and the opamp being connected as inverting amplifier, (3.130) reduces to

$$V_i = -Au_i \tag{3.131}$$

This relation between the internal and output states of the ith neuron shall prove useful while understanding the convergence properties of the network. We shall consider a 2 variable linear equation solver, comprising of 2 neuronal amplifiers and 2 synaptic comparators, to provide an insight into the dynamics of the network. Recalling that the application of KCL for the first neuron resulted in

$$C_{p1}\frac{du_1}{dt} = \frac{V_m \tanh \beta(a_{11}V_1 + a_{12}V_2 - b_1)}{R_{11}}$$
$$+ \frac{V_m \tanh \beta(a_{21}V_1 + a_{22}V_2 - b_2)}{R_{21}} - \frac{u_1}{R_{eff1}} \quad (3.132)$$

Using (3.131) in (3.132), we have

$$C_{p1}\frac{1}{A}\frac{dV_1}{dt} = \frac{V_m \tanh \beta(a_{11}V_1 + a_{12}V_2 - b_1)}{R_{11}}$$
$$+ \frac{V_m \tanh \beta(a_{21}V_1 + a_{22}V_2 - b_2)}{R_{21}} - \frac{1}{A}\frac{V_1}{R_{eff1}} \quad (3.133)$$

Rearranging (3.133) to get terms containing 'time' on one side, we obtain

$$dt = dV_1 \frac{-C_{p1}}{D} \quad (3.134)$$

where the denominator D is

$$D = \frac{A}{R_{11}} V_m \tanh \beta(a_{11}V_1 + a_{12}V_2 - b_1)$$
$$+ \frac{A}{R_{21}} V_m \tanh \beta(a_{21}V_1 + a_{22}V_2 - b_2) + \frac{V_1}{R_{eff1}} \quad (3.135)$$

At this juncture, we shall make a simplifying assumption. For the comparators, the open-loop gain is practically very high ($\beta \to \infty$), and the $\tanh \beta(.)$ remains constant for all practical purposes. Therefore, (3.134) may be simplified as

$$dt = dV_1 \frac{-C_{p1}}{K + \frac{V_1}{R_{eff1}}} \quad (3.136)$$

where

$$K = A V_m [\tanh \beta(a_{11}V_1 + a_{12}V_2 - b_1) + \tanh \beta(a_{21}V_1 + a_{22}V_2 - b_2)] \quad (3.137)$$

Integrating (3.136), we get the following expression for convergence time:

$$T = t_{final} - t_o \quad (3.138)$$

$$T = C_{p1}R_{eff1} \ln \left[\frac{K + \frac{V_i(o)}{R_{eff1}}}{K + \frac{V_i(final)}{R_{eff1}}} \right] \quad (3.139)$$

Also, at the steady state, the output voltage of a neuron no longer changes. Therefore,

$$\frac{dV_1}{dt} = 0 \tag{3.140}$$

Equation (3.140) used in (3.136) yields

$$K + \frac{V_1}{R_{eff1}} = 0 \tag{3.141}$$

which gives

$$V_1 = -K R_{eff1} \tag{3.142}$$

Another noteworthy point is that the values of $V_1(t)$ change from $V_1(0)$ to the steady state value $V_1(\infty)$ in an asymptotic manner. It is advisable, therefore, to measure the convergence time as the time taken to reach a certain fraction of the asymptote. Similarly, we assume that the time is measured from the point when $V_1(t)$ has reached a fraction $\rho_{(0)}$ of the asymptote.
Mathematically,

$$V_1(final) = -\rho_T K R_{eff1} \tag{3.143}$$
$$V_1(0) = -\rho_o K R_{eff1} \tag{3.144}$$

Using (3.143) and (3.144) in (3.139), we get

$$T = C_{p1} \bullet R_{eff1} \bullet \ln\left[\frac{1 - \rho_o}{1 - \rho_T}\right] \tag{3.145}$$

which indicates that the time taken is independent of the exact answer i.e., of the final value at the output node of the first neuron. A similar result can be proved for the second neuron too. In other words, all the nodes converge to a fraction ρ_T of their asymptote at the same time.

An approximate value of T can be estimated by assuming the initial neuron state to be zero i.e, $\rho_o = 0$, and then finding the time taken to reach 99.5 % of the asymptote, we have

$$T = C_{p1} \bullet R_{eff1} \bullet \ln\left[\frac{1 - 0}{1 - 0.995}\right] \tag{3.146}$$

which gives

$$T = C_{p1} \bullet R_{eff1} \bullet \ln(200) \tag{3.147}$$

From the above discussion, following observations are to be noted:

- The convergence time in *not* dependent on V_m—the saturating limit of the comparator
- The convergence time is *not* dependent on n—the number of neurons
- The convergence time is independent of the final values of the neuron states

3.4 Hardware Simulation Results

The NOSYNN-based circuit was tested using PSPICE simulation program for solving sets of 2, 3, 4, 5, 10 and 20 simultaneous linear equations. As has been the approach followed in this text, we shall first consider a simple circuit for solving 2 linear equations in 2 variables. Once the calculation of resistance values for the circuit is explained, the results for the said circuit will be discussed. Thereafter, we shall proceed to problems in 3 (and more) variables.

PSPICE Simulation: 2 variable problem
A simple circuit comprising of 2 neurons (realized using operational amplifiers) and 2 non-linear synapses (comprising of comparators which are also realized using opamps) was first tested using computer simulations using the PSPICE program. Consider the following system of linear equations in 2 variables:

$$\begin{bmatrix} 2 & 3 \\ 6 & 4 \end{bmatrix} \begin{bmatrix} V_1 \\ V_2 \end{bmatrix} = \begin{bmatrix} 4 \\ 2 \end{bmatrix} \tag{3.148}$$

The mathematical solution of (3.148) is $V_1 = -1$ and $V_2 = +2$. Selecting $K = 12\,K\Omega$, the values of the resistances required in synaptic interconnections can be found as

$$R_{11} = \frac{K}{a_{11}} = \frac{12\,K\Omega}{2} = 6\,K\Omega$$
$$R_{12} = \frac{K}{a_{12}} = \frac{12\,K\Omega}{3} = 4\,K\Omega$$
$$R_{21} = \frac{K}{a_{21}} = \frac{12\,K\Omega}{6} = 2\,K\Omega$$
$$R_{22} = \frac{K}{a_{22}} = \frac{12\,K\Omega}{4} = 3\,K\Omega$$

To find the values of resistance required at the input of the comparators, the two equations need to be properly scaled. Choosing a scaling factor of 10 for both the equations, we can write the two equations as

$$\frac{2}{10}V_1 + \frac{3}{10}V_2 = \frac{4}{10} \tag{3.149}$$
$$\frac{6}{10}V_1 + \frac{4}{10}V_2 = \frac{2}{10} \tag{3.150}$$

From (3.149) and (3.150), the resistances that are needed to obtain the required inputs to the two comparators can be found as

$$R_{e11} = 5\,K\Omega$$
$$R_{e12} = 3.33\,K\Omega$$

$$R_{e13} = 2\,K\Omega$$
$$R_{e21} = 1.66\,K\Omega$$
$$R_{e22} = 2.5\,K\Omega$$
$$R_{e23} = \infty$$

and the voltages that need to be applied at the inverting inputs of the two comparators are now given by

$$b_1 = 0.4\,V$$
$$b_2 = 0.2\,V$$

A netlist for the 2 neuron circuit with the above calculated values of resistances was prepared and run in PSPICE. Operational amplifiers were modelled by the μA741 opamp model already available in SPICE. Biasing voltages of the opamp were kept at $\pm 15\,V$. Node voltages were randomly initialized to values in milli-volt range to emulate 'power-on' noise effects. The obtained steady state values of the neuronal voltages were $V_1 = -1.01\,V$ and $V_2 = +2.00\,V$, which are in close proximity with the algebraic solution.

PSPICE Simulation: 3 variable problem

Next, the NOSYNN-based linear equation solver was set up to solve the 3—variable problem given in (3.151). Since the number of decision variables is three, the linear equations solver would comprise of 3 neurons and 3 non-linear synapses (comparators) along with 9 synaptic resistances and 9–12 resistances for obtaining the voltage to be applied at the inputs of the comparators.

$$\begin{bmatrix} 2 & 1 & 1 \\ 3 & 2 & 1 \\ 1 & 1 & 2 \end{bmatrix} \begin{bmatrix} V_1 \\ V_2 \\ V_3 \end{bmatrix} = \begin{bmatrix} 5 \\ 10 \\ 6 \end{bmatrix} \tag{3.151}$$

A circuit to solve 3 variable problems of the type (3.151), as obtained from Fig. 3.8, has already been presented in Fig. 3.9. An analysis similar to that done for the case of the 2 variable circuit needs to be performed for obtaining the values of the various resistances used.

Selecting $K = 6\,K\Omega$, the values of the resistances required in synaptic interconnections can be found as

$$R_{11} = \frac{K}{a_{11}} = \frac{6\,K\Omega}{2} = 3\,K\Omega$$
$$R_{12} = \frac{K}{a_{12}} = \frac{6\,K\Omega}{1} = 6\,K\Omega$$
$$R_{13} = \frac{K}{a_{13}} = \frac{6\,K\Omega}{6} = 6\,K\Omega$$

$$R_{21} = \frac{K}{a_{21}} = \frac{6\,K\Omega}{3} = 2\,K\Omega$$

$$R_{22} = \frac{K}{a_{22}} = \frac{6\,K\Omega}{2} = 3\,K\Omega$$

$$R_{23} = \frac{K}{a_{23}} = \frac{6\,K\Omega}{1} = 6\,K\Omega$$

$$R_{31} = \frac{K}{a_{31}} = \frac{6\,K\Omega}{1} = 6\,K\Omega$$

$$R_{32} = \frac{K}{a_{32}} = \frac{6\,K\Omega}{1} = 6\,K\Omega$$

$$R_{33} = \frac{K}{a_{33}} = \frac{6\,K\Omega}{2} = 3\,K\Omega$$

To find the values of resistance required at the input of the comparators, the three equations need to be properly scaled. Choosing a scaling factor of 6 for all the equations, we can write the scaled version of the three equations as

$$\frac{2}{6}V_1 + \frac{1}{6}V_2 + \frac{1}{6}V_3 = \frac{5}{6} \tag{3.152}$$

$$\frac{3}{6}V_1 + \frac{2}{6}V_2 + \frac{1}{6}V_3 = \frac{10}{6} \tag{3.153}$$

$$\frac{1}{6}V_1 + \frac{1}{6}V_2 + \frac{2}{6}V_3 = \frac{6}{6} \tag{3.154}$$

From (3.152) through (3.154), the resistances that are needed to obtain the required inputs to the three comparators can be found as

$$\frac{1}{R_{e11}} = \frac{a_{11}}{s_1} \Rightarrow R_{e11} = 3\,K\Omega$$

$$\frac{1}{R_{e12}} = \frac{a_{12}}{s_1} \Rightarrow R_{e12} = 6\,K\Omega$$

$$\frac{1}{R_{e13}} = \frac{a_{13}}{s_1} \Rightarrow R_{e13} = 6\,K\Omega$$

$$\frac{1}{R_{e14}} = \left[1 - \left(\frac{1}{R_{e11}} + \frac{1}{R_{e12}} + \frac{1}{R_{e13}}\right)\right] \Rightarrow R_{e14} = 3\,K\Omega$$

$$\frac{1}{R_{e21}} = \frac{a_{21}}{s_2} \Rightarrow R_{e21} = 2\,K\Omega$$

$$\frac{1}{R_{e22}} = \frac{a_{22}}{s_2} \Rightarrow R_{e22} = 3\,K\Omega$$

$$\frac{1}{R_{e23}} = \frac{a_{23}}{s_2} \Rightarrow R_{e23} = 6\,K\Omega$$

$$\frac{1}{R_{e24}} = \left[1 - \left(\frac{1}{R_{e21}} + \frac{1}{R_{e22}} + \frac{1}{R_{e23}}\right)\right] \Rightarrow R_{e24} = \infty$$

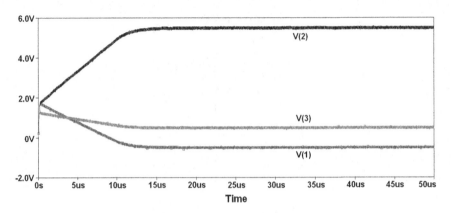

Fig. 3.11 Simulation results for the chosen 3—variable problem

$$\frac{1}{R_{e31}} = \frac{a_{31}}{s_3} \Rightarrow R_{e31} = 6\,K\Omega$$

$$\frac{1}{R_{e32}} = \frac{a_{32}}{s_3} \Rightarrow R_{e32} = 6\,K\Omega$$

$$\frac{1}{R_{e33}} = \frac{a_{33}}{s_3} \Rightarrow R_{e33} = 3\,K\Omega$$

$$\frac{1}{R_{e34}} = \left[1 - \left(\frac{1}{R_{e31}} + \frac{1}{R_{e32}} + \frac{1}{R_{e33}}\right)\right] \Rightarrow R_{e34} = 3\,K\Omega$$

and after scaling, the voltages that need to be applied at the inverting inputs of the three comparators are given by

$$b_1 = 0.833\,V$$
$$b_2 = 1.66\,V$$
$$b_3 = 1\,V$$

Algebraic analysis of (3.151) gives the solution as $V_1 = -0.5$ V, $V_2 = 5.5$ V and $V_3 = 0.5$ V. The results of PSPICE simulation of the circuit of Fig. 3.9, shown in Fig. 3.11, are found to match perfectly with the algebraic solution. The initial node voltages were kept as $V(1) = 10\,mV$, $V(2) = -1\,mV$ and $V(3) = 20\,mV$. For the purpose of this simulation, the LM318 opamp from the Orcad library in PSPICE was utilised. The transfer characteristics of the opamp are presented in Fig. 3.12. The value of β for this opamp was measured to be 1.1×10^4 using PSPICE simulation. The NOSYNN-based circuit was further tested in PSPICE for solving systems of linear equations in 4, 5, 10 and 20 variables. Results of simulation runs for these problems are presented in Table 3.1. Each of the simulations was run using various initial conditions in the millivolt range. As can be seen from Table 3.1, the NOSYNN-based network always converges to the solution of the given system of linear equations.

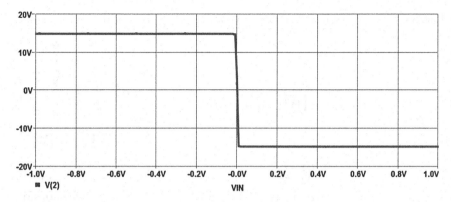

Fig. 3.12 Transfer characteristics of the opamps used to realize the neurons of Fig. 3.9

3.5 Hardware Implementation

Breadboard implementations of the NOSYNN based circuit yielded promising results. Apart from verification of the working of the NOSYNN-based linear equation solver, the actual hardware realization also served the purpose of testing the convergence of the circuit to the solution starting from different initial conditions. The noise present in any electronic circuit acts as a random initial condition for the convergence of the neural circuit. Standard laboratory components *viz.* the μA741 opamp and resistances were used for the purpose. Circuits to solve sets of equations in 2 and 3 variables were assembled and each circuit was tested for 25 test runs. The results of hardware implementation are presented in Table 3.2. Snapshots of the experimental setups for the *first* sets of 2—and 3—variable problems of Table 3.2 are presented in Figs. 3.13 and 3.14 respectively. It can be seen that the network converged to the correct solution in all the test cases, and the deviation from the algebraic solution was low for every system of linear equation.

It needs to be mentioned that the μA741 operational amplifier was used for breadboard implementations due to its easy availability, low price and near ideal amplifier characteristics. If a different opamp is selected for setting up the circuits, the results in Table 3.2 are not expected to change much. Another important point to consider while attempting such hardware realizations using discrete resistors is the tolerance values associated with them. A typical laboratory resistor would have a value in the range of ±5–10 % of its nominal value. This would cause errors in the obtained solutions when a linear equation solver is configured in a laboratory. Readers who are interested in testing such circuits in hardware should therefore find ways to circumvent the problem of tolerances in the values of discrete resistors. One viable option is to measure a resistor (instead of following its colour code) using a multimeter and add an adjustment resistor in case the value deviates from the nominal value. It was this approach which ensured the high solution quality depicted in Table 3.2.

Table 3.1 PSPICE simulation results for the NOSYNN-based circuit applied to various problems

[A]	[B]	Algebraic Solution [V]	Simulated Results (Using PSPICE) [V]
$\begin{bmatrix} 1 & 2 \\ 2 & 1 \end{bmatrix}$	$\begin{bmatrix} 3.5 \\ 4 \end{bmatrix}$	$\begin{bmatrix} 1.5 \\ 1 \end{bmatrix}$	$\begin{bmatrix} 1.51 \\ 1.02 \end{bmatrix}$
$\begin{bmatrix} 2 & 1 & 1 \\ 3 & 2 & 1 \\ 1 & 1 & 2 \end{bmatrix}$	$\begin{bmatrix} 5 \\ 10 \\ 6 \end{bmatrix}$	$\begin{bmatrix} -0.5 \\ 5.5 \\ 0.5 \end{bmatrix}$	$\begin{bmatrix} -0.50 \\ 5.51 \\ 0.51 \end{bmatrix}$
$\begin{bmatrix} 2 & 3 & 6 & 1 \\ 3 & 2 & 5 & 4 \\ 2 & 5 & 5 & 2 \\ 2 & 4 & 2 & 5 \end{bmatrix}$	$\begin{bmatrix} 48.75 \\ 67.5 \\ 54 \\ 59.25 \end{bmatrix}$	$\begin{bmatrix} 3.5 \\ 1.5 \\ 5 \\ 7.25 \end{bmatrix}$	$\begin{bmatrix} 3.57 \\ 1.51 \\ 4.92 \\ 7.20 \end{bmatrix}$
$\begin{bmatrix} 2 & 3 & 9 & 2 & 5 \\ 2 & 6 & 9 & 9 & 5 \\ 2 & 6 & 2 & 4 & 5 \\ 2 & 4 & 7 & 8 & 3 \\ 5 & 3 & 6 & 3 & 5 \end{bmatrix}$	$\begin{bmatrix} -54.9 \\ -91.0 \\ -54.9 \\ -74.8 \\ -70.0 \end{bmatrix}$	$\begin{bmatrix} -6.7 \\ -4.3 \\ -2.8 \\ -3.3 \\ 0.64 \end{bmatrix}$	$\begin{bmatrix} -6.62 \\ -4.15 \\ -2.65 \\ -3.18 \\ 0.58 \end{bmatrix}$
$\begin{bmatrix} 1 & 2 & 1 & 3 & 4 & 2 & 1 & 1 & 1 & 2 \\ 2 & 1 & 1 & 3 & 1 & 2 & 2 & 1 & 2 & 3 \\ 1 & 1 & 4 & 3 & 3 & 1 & 1 & 4 & 4 & 1 \\ 4 & 2 & 1 & 5 & 3 & 3 & 1 & 1 & 2 & 2 \\ 1 & 1 & 5 & 1 & 2 & 1 & 2 & 2 & 5 & 2 \\ 3 & 3 & 1 & 2 & 1 & 1 & 5 & 2 & 1 & 1 \\ 5 & 5 & 1 & 4 & 1 & 1 & 3 & 4 & 2 & 2 \\ 1 & 1 & 1 & 2 & 2 & 2 & 3 & 3 & 4 & 1 \\ 4 & 2 & 2 & 1 & 3 & 2 & 5 & 4 & 3 & 2 \\ 1 & 2 & 3 & 1 & 2 & 3 & 1 & 2 & 3 & 4 \end{bmatrix}$	$\begin{bmatrix} 10 \\ -11 \\ 10 \\ 2 \\ -7 \\ -9 \\ -8 \\ -3 \\ -3 \\ 7 \end{bmatrix}$	$\begin{bmatrix} -2 \\ 1 \\ 3 \\ -1 \\ 2 \\ 7 \\ -3 \\ 4 \\ -5 \\ -4 \end{bmatrix}$	$\begin{bmatrix} -1.99 \\ 1.00 \\ 3.00 \\ -1.00 \\ 1.99 \\ 7.00 \\ -3.00 \\ 3.99 \\ -5.00 \\ -4.00 \end{bmatrix}$
$\begin{bmatrix} 7 & 2 & 3 & 5 & 4 & 7 & 2 & 7 & 7 & 1 & 3 & 5 & 2 & 2 & 6 & 7 & 7 & 1 & 2 & 7 \\ 3 & 3 & 1 & 6 & 6 & 7 & 6 & 1 & 6 & 3 & 6 & 5 & 2 & 7 & 3 & 4 & 7 & 3 & 1 \\ 3 & 5 & 2 & 4 & 5 & 7 & 4 & 5 & 3 & 6 & 5 & 4 & 4 & 6 & 5 & 3 & 3 & 5 & 5 & 2 \\ 5 & 3 & 7 & 2 & 6 & 5 & 2 & 2 & 4 & 4 & 3 & 7 & 6 & 6 & 7 & 1 & 2 & 7 & 7 & 7 \\ 7 & 6 & 7 & 5 & 2 & 1 & 7 & 5 & 5 & 5 & 2 & 4 & 2 & 7 & 5 & 6 & 5 & 7 & 3 & 3 \\ 1 & 2 & 1 & 1 & 2 & 3 & 7 & 5 & 7 & 2 & 3 & 4 & 6 & 3 & 5 & 7 & 6 & 7 & 2 \\ 1 & 6 & 3 & 2 & 1 & 1 & 7 & 5 & 3 & 6 & 7 & 4 & 6 & 4 & 7 & 2 & 4 & 7 & 2 & 3 \\ 2 & 2 & 1 & 1 & 1 & 2 & 3 & 5 & 6 & 2 & 2 & 4 & 2 & 1 & 3 & 5 & 4 & 3 & 4 & 5 \\ 4 & 4 & 2 & 2 & 3 & 1 & 4 & 3 & 3 & 4 & 2 & 5 & 1 & 5 & 3 & 6 & 2 & 1 & 4 & 3 \\ 1 & 6 & 1 & 3 & 2 & 2 & 2 & 3 & 6 & 3 & 6 & 4 & 7 & 1 & 1 & 5 & 2 & 5 & 7 & 3 \\ 6 & 2 & 3 & 3 & 1 & 5 & 4 & 4 & 5 & 6 & 6 & 5 & 3 & 7 & 7 & 7 & 7 & 7 & 4 & 7 \\ 1 & 1 & 2 & 1 & 4 & 5 & 1 & 5 & 6 & 1 & 5 & 2 & 7 & 2 & 4 & 2 & 2 & 3 & 3 & 6 \\ 2 & 7 & 5 & 3 & 5 & 4 & 7 & 4 & 2 & 4 & 3 & 5 & 6 & 1 & 6 & 2 & 7 & 1 & 3 & 7 \\ 1 & 3 & 5 & 3 & 7 & 7 & 2 & 2 & 1 & 2 & 7 & 4 & 2 & 4 & 4 & 7 & 5 & 1 & 1 & 1 \\ 3 & 2 & 6 & 4 & 1 & 3 & 5 & 3 & 4 & 6 & 7 & 4 & 3 & 2 & 2 & 1 & 7 & 1 & 5 & 7 \\ 5 & 4 & 7 & 2 & 1 & 6 & 1 & 3 & 4 & 3 & 3 & 3 & 6 & 1 & 2 & 1 & 3 & 6 & 7 & 4 \\ 2 & 4 & 2 & 6 & 4 & 1 & 7 & 3 & 5 & 3 & 5 & 5 & 4 & 1 & 3 & 6 & 5 & 4 & 2 & 5 \\ 5 & 3 & 6 & 4 & 3 & 1 & 5 & 6 & 7 & 4 & 4 & 7 & 1 & 3 & 1 & 2 & 7 & 6 & 6 & 5 \\ 2 & 4 & 5 & 1 & 7 & 1 & 3 & 3 & 6 & 4 & 6 & 4 & 5 & 6 & 7 & 3 & 7 & 1 & 2 & 4 \\ 3 & 5 & 2 & 3 & 5 & 3 & 5 & 2 & 4 & 5 & 4 & 7 & 3 & 6 & 6 & 2 & 3 & 4 & 6 & 7 \end{bmatrix}$	$\begin{bmatrix} 84.9 \\ 108.3 \\ 66 \\ -11.2 \\ 89.5 \\ 23.7 \\ 72.8 \\ 16.7 \\ 73.2 \\ -22.1 \\ 98.8 \\ -43.9 \\ 56.4 \\ 124.6 \\ 49.8 \\ -35.7 \\ 73.2 \\ 42.9 \\ 47.3 \\ 31.9 \end{bmatrix}$	$\begin{bmatrix} 2 \\ -1 \\ 0 \\ 3.5 \\ -1.5 \\ 1 \\ 1 \\ 2.5 \\ -7 \\ 6.5 \\ 3 \\ 9 \\ -4 \\ 0 \\ 2.4 \\ 6 \\ 4 \\ -2 \\ -8 \\ -5 \end{bmatrix}$	$\begin{bmatrix} 2.19 \\ -1.05 \\ -0.10 \\ 3.30 \\ -1.19 \\ 0.94 \\ 1.23 \\ 2.78 \\ -7.21 \\ 6.23 \\ 3.22 \\ 8.52 \\ -3.99 \\ 0.15 \\ 2.16 \\ 5.98 \\ 3.87 \\ -1.92 \\ -7.89 \\ -4.71 \end{bmatrix}$

Table 3.2 Hardware test results of the NOSYNN-based circuit applied to 2 and 3 variable problems

Chosen Set of Linear Equations	Algebraic Solution	Hardware Test Results (First 5 runs)					Standard Deviation, σ (for 25 runs)
$\begin{bmatrix}1&1\\1&3\end{bmatrix}\begin{bmatrix}V_1\\V_2\end{bmatrix}=\begin{bmatrix}10\\10\end{bmatrix}$	$\begin{bmatrix}V_1\\V_2\end{bmatrix}=\begin{bmatrix}10\\0\end{bmatrix}$	$\begin{bmatrix}9.89\\0.18\end{bmatrix}$	$\begin{bmatrix}9.90\\0.18\end{bmatrix}$	$\begin{bmatrix}9.89\\0.18\end{bmatrix}$	$\begin{bmatrix}9.89\\0.18\end{bmatrix}$	$\begin{bmatrix}9.89\\0.18\end{bmatrix}$	$\begin{bmatrix}\sigma_1\\\sigma_2\end{bmatrix}=\begin{bmatrix}0.039\\0.017\end{bmatrix}$
$\begin{bmatrix}2&1\\1&3\end{bmatrix}\begin{bmatrix}V_1\\V_2\end{bmatrix}=\begin{bmatrix}8\\-1\end{bmatrix}$	$\begin{bmatrix}V_1\\V_2\end{bmatrix}=\begin{bmatrix}5\\-2\end{bmatrix}$	$\begin{bmatrix}4.97\\-2.0\end{bmatrix}$	$\begin{bmatrix}4.97\\-2.0\end{bmatrix}$	$\begin{bmatrix}4.98\\-1.99\end{bmatrix}$	$\begin{bmatrix}4.97\\-2.0\end{bmatrix}$	$\begin{bmatrix}4.98\\-2.0\end{bmatrix}$	$\begin{bmatrix}\sigma_1\\\sigma_2\end{bmatrix}=\begin{bmatrix}0.021\\0.005\end{bmatrix}$
$\begin{bmatrix}5&3\\4&6\end{bmatrix}\begin{bmatrix}V_1\\V_2\end{bmatrix}=\begin{bmatrix}11\\16\end{bmatrix}$	$\begin{bmatrix}V_1\\V_2\end{bmatrix}=\begin{bmatrix}1\\2\end{bmatrix}$	$\begin{bmatrix}1.01\\1.98\end{bmatrix}$	$\begin{bmatrix}1.01\\1.99\end{bmatrix}$	$\begin{bmatrix}1.01\\1.98\end{bmatrix}$	$\begin{bmatrix}1.00\\1.97\end{bmatrix}$	$\begin{bmatrix}1.01\\1.98\end{bmatrix}$	$\begin{bmatrix}\sigma_1\\\sigma_2\end{bmatrix}=\begin{bmatrix}0.028\\0.037\end{bmatrix}$
$\begin{bmatrix}4&1\\1&4\end{bmatrix}\begin{bmatrix}V_1\\V_2\end{bmatrix}=\begin{bmatrix}1\\4\end{bmatrix}$	$\begin{bmatrix}V_1\\V_2\end{bmatrix}=\begin{bmatrix}0\\1\end{bmatrix}$	$\begin{bmatrix}0.01\\1.01\end{bmatrix}$	$\begin{bmatrix}0.02\\0.99\end{bmatrix}$	$\begin{bmatrix}0.01\\0.99\end{bmatrix}$	$\begin{bmatrix}0.01\\0.99\end{bmatrix}$	$\begin{bmatrix}0.02\\0.99\end{bmatrix}$	$\begin{bmatrix}\sigma_1\\\sigma_2\end{bmatrix}=\begin{bmatrix}0.032\\0.027\end{bmatrix}$
$\begin{bmatrix}2&1\\3&4\end{bmatrix}\begin{bmatrix}V_1\\V_2\end{bmatrix}=\begin{bmatrix}8\\17\end{bmatrix}$	$\begin{bmatrix}V_1\\V_2\end{bmatrix}=\begin{bmatrix}3\\2\end{bmatrix}$	$\begin{bmatrix}3.01\\2.01\end{bmatrix}$	$\begin{bmatrix}3.01\\2.00\end{bmatrix}$	$\begin{bmatrix}3.01\\2.01\end{bmatrix}$	$\begin{bmatrix}3.00\\2.01\end{bmatrix}$	$\begin{bmatrix}3.01\\2.01\end{bmatrix}$	$\begin{bmatrix}\sigma_1\\\sigma_2\end{bmatrix}=\begin{bmatrix}0.018\\0.006\end{bmatrix}$
$\begin{bmatrix}1&1&3\\1&3&1\\3&1&1\end{bmatrix}\begin{bmatrix}V_1\\V_2\\V_3\end{bmatrix}=\begin{bmatrix}10\\10\\10\end{bmatrix}$	$\begin{bmatrix}V_1\\V_2\\V_3\end{bmatrix}=\begin{bmatrix}2\\2\\2\end{bmatrix}$	$\begin{bmatrix}2.04\\1.98\\2.00\end{bmatrix}$	$\begin{bmatrix}2.01\\1.97\\2.00\end{bmatrix}$	$\begin{bmatrix}2.04\\1.98\\2.01\end{bmatrix}$	$\begin{bmatrix}2.04\\1.98\\2.00\end{bmatrix}$	$\begin{bmatrix}2.04\\1.97\\2.01\end{bmatrix}$	$\begin{bmatrix}\sigma_1\\\sigma_2\\\sigma_3\end{bmatrix}=\begin{bmatrix}0.044\\0.025\\0.032\end{bmatrix}$
$\begin{bmatrix}2&3&4\\4&3&2\\2&4&3\end{bmatrix}\begin{bmatrix}V_1\\V_2\\V_3\end{bmatrix}=\begin{bmatrix}20\\16\\19\end{bmatrix}$	$\begin{bmatrix}V_1\\V_2\\V_3\end{bmatrix}=\begin{bmatrix}1\\2\\3\end{bmatrix}$	$\begin{bmatrix}1.03\\1.97\\2.99\end{bmatrix}$	$\begin{bmatrix}1.03\\1.97\\2.99\end{bmatrix}$	$\begin{bmatrix}1.03\\1.97\\2.97\end{bmatrix}$	$\begin{bmatrix}1.04\\1.98\\2.98\end{bmatrix}$	$\begin{bmatrix}1.03\\1.97\\2.99\end{bmatrix}$	$\begin{bmatrix}\sigma_1\\\sigma_2\\\sigma_3\end{bmatrix}=\begin{bmatrix}0.034\\0.035\\0.022\end{bmatrix}$
$\begin{bmatrix}1&3&2\\2&1&1\\1&2&2\end{bmatrix}\begin{bmatrix}V_1\\V_2\\V_3\end{bmatrix}=\begin{bmatrix}3\\3\\3\end{bmatrix}$	$\begin{bmatrix}V_1\\V_2\\V_3\end{bmatrix}=\begin{bmatrix}1\\0\\1\end{bmatrix}$	$\begin{bmatrix}1.00\\0.02\\1.01\end{bmatrix}$	$\begin{bmatrix}0.98\\0.02\\1.01\end{bmatrix}$	$\begin{bmatrix}0.99\\0.01\\1.02\end{bmatrix}$	$\begin{bmatrix}0.99\\0.02\\1.01\end{bmatrix}$	$\begin{bmatrix}0.99\\0.01\\1.01\end{bmatrix}$	$\begin{bmatrix}\sigma_1\\\sigma_2\\\sigma_3\end{bmatrix}=\begin{bmatrix}0.024\\0.025\\0.022\end{bmatrix}$
$\begin{bmatrix}2&2&2\\3&4&1\\1&2&3\end{bmatrix}\begin{bmatrix}V_1\\V_2\\V_3\end{bmatrix}=\begin{bmatrix}6\\10\\4\end{bmatrix}$	$\begin{bmatrix}V_1\\V_2\\V_3\end{bmatrix}=\begin{bmatrix}2\\1\\0\end{bmatrix}$	$\begin{bmatrix}1.96\\1.00\\0.01\end{bmatrix}$	$\begin{bmatrix}1.98\\1.01\\0.01\end{bmatrix}$	$\begin{bmatrix}1.98\\1.01\\0.01\end{bmatrix}$	$\begin{bmatrix}1.96\\1.02\\0.01\end{bmatrix}$	$\begin{bmatrix}1.96\\1.01\\0.02\end{bmatrix}$	$\begin{bmatrix}\sigma_1\\\sigma_2\\\sigma_3\end{bmatrix}=\begin{bmatrix}0.054\\0.045\\0.037\end{bmatrix}$
$\begin{bmatrix}1&2&3\\4&2&6\\6&8&2\end{bmatrix}\begin{bmatrix}V_1\\V_2\\V_3\end{bmatrix}=\begin{bmatrix}1\\10\\-4\end{bmatrix}$	$\begin{bmatrix}V_1\\V_2\\V_3\end{bmatrix}=\begin{bmatrix}2\\-2\\1\end{bmatrix}$	$\begin{bmatrix}2.02\\-1.99\\1.02\end{bmatrix}$	$\begin{bmatrix}2.00\\-1.99\\1.02\end{bmatrix}$	$\begin{bmatrix}2.02\\-1.98\\1.01\end{bmatrix}$	$\begin{bmatrix}2.02\\-1.98\\1.00\end{bmatrix}$	$\begin{bmatrix}2.02\\-2.01\\0.99\end{bmatrix}$	$\begin{bmatrix}\sigma_1\\\sigma_2\\\sigma_3\end{bmatrix}=\begin{bmatrix}0.029\\0.035\\0.019\end{bmatrix}$

3.6 Low-Voltage CMOS-Compatible Linear Equation Solver

The results of PSPICE simulation as well as experimental verification of the NOSYNN-based voltage-mode neural circuit for solving linear equations, presented in the previous sections, are in excellent agreement with the algebraic solutions for all the sets of simultaneous linear equations considered. However, both the operational amplifiers used (LM318 for PSPICE simulations and μA741 for breadboard implementations) are designed in bipolar technology [15, 16].

Since modern day integration processes tend to favour CMOS technology, it is prudent to test the operation of the linear equation solver using a CMOS operational amplifier in sub-micron technology. Further, since low voltage operation is

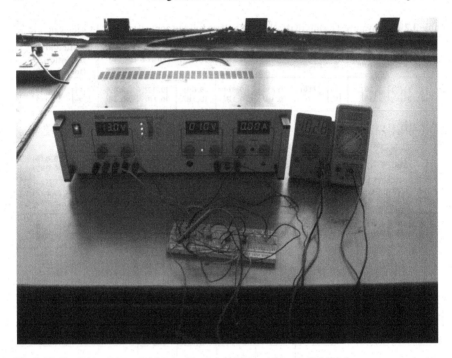

Fig. 3.13 Experimental results for the first 2—variable problem in Table 3.2

another desirable feature in the design of any electronic circuit, a low-voltage CMOS-compatible operational amplifier was chosen for use in the NOSYNN-based circuit. The low-voltage CMOS rail-to-rail operational amplifier using double p-channel differential input pairs, taken from [17], is capable of satisfactory operation using ± 0.9 V power supplies. The amplifier boasts of a DC gain of 72.2 dB, a CMRR of 67.2 dB and power dissipation of 288 μW[17].

The NOSYNN-based neural circuit was tested using PSPICE simulations for solving systems of linear equations of various sizes ranging from 2—to 10—variables. For the opamps, biasing supplies were kept at ± 0.9 V and 0.35μm CMOS n-well process device models were used for PSPICE simulation. The sample sets of linear equations and the results as obtained after circuit simulations are presented in Table 3.3 from where it is evident that the NOSYNN-based circuit is able to converge to the correct solution for all cases and the obtained solutions are in close agreement with the algebraic solutions. The results of PSPICE simulation as obtained for the 10—variable problem in Table 3.3 are presented in Fig. 3.15 from where it can be observed that the neuron states settle to voltages which correspond to the solution points of the 10—variable system of equations.

It may be noted that the CMOS operational amplifier used above is one of the numerous CMOS-compatible amplifiers available in the technical literature, and was chosen in view of its various attractive features like high DC gain and high CMRR

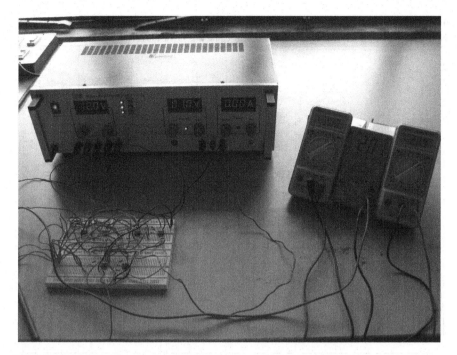

Fig. 3.14 Experimental results for the first 3—variable problem in Table 3.2

at low biasing voltages. However, since the appearance of the said opamp, a lot of other better performing amplifiers have been added to the already extensive pool of opamps. A constant-g_m CMOS op-amp with rail-to-rail input and output stages working at $1.5\,V$ and having a DC gain of $106.6\,dB$ has been reported [18]. An ultra low power low voltage high CMRR differential amplifier with rail-to-rail input common mode range based on quasi floating gate transistors has also been proposed [19]. The required supply voltage is only ($V_{GS} + V_{DSsat}$) which is one V_{DSsat} lower than the required supply voltage for conventional differential amplifiers. The circuit is designed using $0.18\,\mu m$ CMOS process parameters and works at $\pm 0.4\,V$ supply voltage. The CMRR, and power dissipation for the opamp are reported to be $121\,dB$ and $6.89\,\mu W$ respectively [19].

The use of a low-voltage opamp, however, has the disadvantage of reducing the range of allowable inputs as well as possible outputs. Linear equations may have to be scaled before application to NOSYNN based circuits employing low-voltage opamps. Since the output of an electronic amplifier is restricted between the highest and lowest biasing voltages, the outputs of all neurons in a linear equation solver implemented with low voltage opamps would be forced to lie in a narrow range (or else, they will saturate) and this may result in problems while retrieving the solution.

Table 3.3 PSPICE simulation results with the low-voltage CMOS opamp employed in the NOSYNN-based circuit to solve system of linear equations of various sizes

[A]	[B]	Algebraic Solution [V]	Simulated Results (Using PSPICE) [V]	Percentage Error in the solution (%)
$\begin{bmatrix} 2 & 3 \\ 4 & 1 \end{bmatrix}$	$\begin{bmatrix} 0.1 \\ 0.7 \end{bmatrix}$	$\begin{bmatrix} 0.2 \\ -0.1 \end{bmatrix}$	$\begin{bmatrix} 200\,mV \\ -99\,mV \end{bmatrix}$	$\begin{bmatrix} 0 \\ -1 \end{bmatrix}$
$\begin{bmatrix} 3 & 2 & 4 \\ 1 & 5 & 2 \\ 3 & 3 & 3 \end{bmatrix}$	$\begin{bmatrix} 1.5 \\ 0.9 \\ 1.0 \end{bmatrix}$	$\begin{bmatrix} -0.193 \\ 0.013 \\ 0.513 \end{bmatrix}$	$\begin{bmatrix} -192\,mV \\ 13\,mV \\ 513\,mV \end{bmatrix}$	$\begin{bmatrix} -0.5 \\ 0 \\ 0 \end{bmatrix}$
$\begin{bmatrix} 1 & 2 & 3 & 4 & 5 \\ 5 & 4 & 3 & 2 & 1 \\ 6 & 1 & 6 & 4 & 2 \\ 3 & 3 & 2 & 4 & 2 \\ 2 & 2 & 5 & 4 & 6 \end{bmatrix}$	$\begin{bmatrix} -7.75 \\ -5.75 \\ -8.05 \\ -6.20 \\ -9.55 \end{bmatrix}$	$\begin{bmatrix} -0.25 \\ -0.35 \\ -0.45 \\ -0.55 \\ -0.65 \end{bmatrix}$	$\begin{bmatrix} -239\,mV \\ -361\,mV \\ -464\,mV \\ -547\,mV \\ -638\,mV \end{bmatrix}$	$\begin{bmatrix} -4.4 \\ 3.1 \\ 3.1 \\ -0.5 \\ -1.8 \end{bmatrix}$
$\begin{bmatrix} 1 & 2 & 1 & 3 & 4 & 2 & 1 & 1 & 1 & 2 \\ 2 & 1 & 1 & 3 & 1 & 2 & 2 & 1 & 2 & 3 \\ 1 & 1 & 4 & 3 & 3 & 1 & 1 & 4 & 4 & 1 \\ 4 & 2 & 1 & 5 & 3 & 3 & 1 & 1 & 2 & 2 \\ 1 & 1 & 5 & 1 & 2 & 1 & 2 & 2 & 5 & 2 \\ 3 & 3 & 1 & 2 & 1 & 1 & 5 & 2 & 1 & 1 \\ 5 & 5 & 1 & 4 & 1 & 1 & 3 & 4 & 2 & 2 \\ 1 & 1 & 1 & 2 & 2 & 2 & 3 & 3 & 4 & 1 \\ 4 & 2 & 2 & 1 & 3 & 2 & 5 & 4 & 3 & 2 \\ 1 & 2 & 3 & 1 & 2 & 3 & 1 & 2 & 3 & 4 \end{bmatrix}$	$\begin{bmatrix} 1 \\ -1.1 \\ 1 \\ 0.2 \\ -0.7 \\ -0.9 \\ -0.8 \\ -0.3 \\ -0.3 \\ 0.7 \end{bmatrix}$	$\begin{bmatrix} -0.2 \\ 0.1 \\ 0.3 \\ -0.1 \\ 0.2 \\ 0.7 \\ -0.3 \\ 0.4 \\ -0.5 \\ -0.4 \end{bmatrix}$	$\begin{bmatrix} -200\,mV \\ 100\,mV \\ 300\,mV \\ -99\,mV \\ 200\,mV \\ 699\,mV \\ -299\,mV \\ 400\,mV \\ -499\,mV \\ -399\,mV \end{bmatrix}$	$\begin{bmatrix} 0 \\ 0 \\ 0 \\ -1 \\ 0 \\ -0.1 \\ -0.3 \\ 0 \\ -0.2 \\ -0.2 \end{bmatrix}$

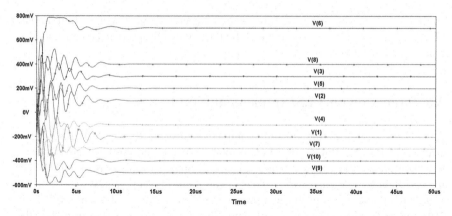

Fig. 3.15 Result of PSPICE simulation as obtained for the 10—variable problem in Table 3.3

3.7 Comparison with Existing Works

The voltage-mode NOSYNN based linear equation solver presented in the present chapter employs $2n$ operational amplifiers and $(2n^2 + n)$ resistances to solve an $n \times n$ system of linear equations. In comparison, Wang's electronic realization of a recurrent neural network for solving linear equations uses $3n$ operational amplifiers, n capacitors and $(n^2 + 5n)$ resistances and has a solution time of the order of hundreds of microseconds [20]. The circuit attributed to Cichocki and Unbehauen is able to provide solution time of the order of microseconds but at the cost of increased hardware complexity requiring $3n$ weighted summers and n inverting integrators [21]. Wang and Li employed a linear activation function in their neural network for solving linear equations [22]. The electronic realization of the network presented in [22] has the same complexity as the one NOSYNN-based in [20]. Xia, Wang and Hung came up with a linear equation solver [23], which is essentially a generalized neural network based implementation of Censor and Elfving's method for linear inequalities [24]. They used an approach similar to [20], utilizing weighted adders and integrators to realize the neurons.

Furthermore, the NOSYNN based circuit for solving linear equations enjoy convergence times of the order of microseconds. For instance, Fig. 3.11 shows that the NOSYNN-based circuit takes approximately 15 μsec to arrive at the solution point for a 3-variable problem. In comparison, the DGESV driver routine in LAPACK (**L**inear **A**lgebra **PACK**age, which provides software routines for solving systems of simultaneous linear equations), has a run-time in milliseconds for solving an $n \times n$ system of linear equations of small problem sizes [25].

While the accuracy of the NOSYNN-based system is not at par with present day digital systems, such analog linear equation solvers may be employed in real-time control applications where long computational times may require that a simpler (and possibly inferior) model be used during the course of design optimization [26].

Such systems may also be used in a symbiotic environment with a digital computer, acting as a co-processor to it thereby improving the speed and convergence of numerical algorithms [27].

3.8 Discussion on VLSI Implementation Issues

This section deals with the monolithic implementation issues of a NOSYNN based circuit. The PSPICE simulations assumed that all operational amplifiers are identical, and therefore, it is necessary to determine how deviations from this assumption affect the performance of the network. Effects of variations in component values from one neuron to another were also investigated using Monte-Carlo analysis in PSPICE. A 10% tolerance with Gaussian deviation profile was put on the resistances used in the circuit of Fig. 3.9. The analysis was carried out for 100 runs and the Mean Deviation was found out to be 0.0497 and Mean Sigma (Standard Deviation) was 0.1243. Offset analysis was also carried out by incorporating random offset voltages (in the range of 1–10 mV) to the opamps. The Mean Deviation in this case was measured to be −0.0512 and the Mean Sigma (Standard Deviation) was 0.172. As can be seen, the effects of mismatches and offsets on the overall precision of the final results are in an acceptable range.

In fact, the drawbacks of the use of opamps in the NOSYNN-based circuit suggest that a real, large scale implementation for solving a system of equations in several variables might be quite different. Alternative realizations based on the differential equations (3.99) governing the system of neurons were also considered. An implementation based on operational transconductance amplifiers (OTAs) would appear to be one viable alternative [28, 29]; similarly, circuits using MOS transistors operating in the sub-threshold regime would be another. In the latter case, the *tanh*(.) non-linearity is easily obtained because of the transistor's current–voltage characteristic [30]. However, the present chapter focuses mainly on the principle of such a network, and details of an actual VLSI implementation are explored in a later chapter.

Further, scaling issues for the NOSYNN-based circuit were also investigated. For this purpose, two problems of 10 and 20 variables each were selected. These appear in Table 3.1 and were solved using the network of Fig. 3.8. PSPICE simulation was performed and the obtained simulation results for the 10-variable problem were $V_1 = -1.99$, $V_2 = 1.00$, $V_3 = 3.00$, $V_4 = -1.00$, $V_5 = 1.99$, $V_6 = 7.00$, $V_7 = -3.00$, $V_8 = 3.99$, $V_9 = -5.00$ and $V_{10} = -4.00$ which agree very closely with the exact mathematical solution.

3.9 Conclusion

A NOSYNN based approach to solve n simultaneous linear equations in n variables, which uses n neurons and n synapses has been presented. Each neuron requires one opamp and each synapse is implemented using one comparator. The energy function associated with the NOSYNN-based linear equation solving circuit has been shown to contain transcendental terms which ensure convergence to the exact solution point. Hardware implementations were carried out for 2 and 3 variable problems. The results were found to match with the mathematical solution. The working of the NOSYNN-based network was also verified using PSPICE for various sample problem sets of 2–20 simultaneous linear equations. From VLSI implementation point of view, a CMOS compatible implementation of the NOSYNN-based circuit is also discussed.

References

1. Xu, Z.-B., Hu, G.-Q., Kwong, C.-P.: Asymmetric hopfield-type networks: theory and applications. Neural Netw. **9**(3), 483–501 (1996)
2. Kosko, B.: Bidirectional associative memories. IEEE Trans. Syst. Man Cybern. **18**(1), 49–60 (1988)
3. Ackley, D.H., Hinton, G.E., Sejnowski, T.J.: A learning algorithm for boltzmann machines. Cogn. Sci. **9**(1), 147–169 (1985)
4. Kohring, G.A.: On the Q-state neuron problem in attractor neural networks. Neural Netw. **6**(4), 573–581 (1993)
5. Guan, Z.-H., Chen, G., Qin, Y.: On equilibria, stability, and instability of hopfield neural networks. IEEE Trans. Neural Netw. **11**(2), 534–540 (Mar 2000)
6. Arik, S.: Global asymptotic stability of a class of dynamical neural networks. IEEE Trans. Circ. Syst. I Fundam. Theory Appl. **47**(4), 568–571 (Apr 2000)
7. Hopfield, J.J., Tank, D.W.: "Neural" computation of decisions in optimization problems. Biol. Cybern. **52**, 141–152 (1985)
8. Hopfield, J.J.: Neural networks and physical systems with emergent collective computational abilities. Proc. Nat. Acad. Sci. **79**(8), 2554–2558 (1982)
9. Hopfield, J.J.: Neurons with graded response have collective computational properties like those of two-state neurons. Proc. Nat. Acad. Sci. **81**(10), 3088–3092 (1984)
10. Vidyasagar, M.: Location and stability of the high-gain equilibria of nonlinear neural networks. IEEE Trans. Neural Netw. **4**(4), 660–672 (1993)
11. Tank, D., Hopfield, J.: Simple 'neural' optimization networks: an A/D converter, signal decision circuit, and a linear programming circuit. IEEE Trans. Circ. Syst. **33**(5), 533–541 (1986)
12. Zurada, J.M.: Introduction to artificial neural systems. West Publishing Co, Saint Paul (1992)
13. Rahman, S.A., Jayadeva, Dutta Roy, S.C.: Neural network approach to graph colouring. Electron. Lett. **35**(14), 1173–1175 (1999)
14. Rahman, S.A.: A nonlinear synapse neural network and its applications. PhD thesis, Department of Electrical Engineering. Indian Institute of Technology, Delhi (2007)
15. LM318 Operational Amplifier. National semiconductor incorporated. online. http://www.national.com/mpf/LM/LM318.html (2012). Accessed 30 Oct 2012
16. μA741 General Purpose Operational Amplifiers. Texas instruments incorporated online. http://www.ti.com/lit/ds/symlink/ua741.pdf (2012) Accessed 30 Oct 2012
17. Huang, C.-J., Huang, H.-Y.: A low-voltage CMOS rail-to-rail operational amplifier using double p-channel differential input pairs. In: Proceedings of the 2004 International Symposium on Circuits and Systems (ISCAS'04), vol. 1, pp. 673–676 (2004)
18. Dai, G.-D., Huang, P., Yang, L., Wang, B.: A constant G_m CMOS op-amp with rail-to-rail input/output stage. In: 10th IEEE International Conference on Solid-State and Integrated Circuit Technology (ICSICT), pp. 123–125 (2010)
19. Safari, L., Azhari, S.J.: An ultra low power, low voltage tailless QFG based differential amplifier with high CMRR, rail to rail operation and enhanced slew rate. Analog Integr. Circ. Sig. Process. **67**(2), 241–252 (2011)

20. Wang, J.: Electronic realization of recurrent neural network for solving simultaneous linear equations. Electron. Lett. **28**(5), 493–495 (1992)
21. Cichocki, A., Unbehauen, R.: Neural networks for solving systems of linear equations and related problems. IEEE Trans. Circ. Syst. I Fundam. Theory Appl. **39**(2), 124–138 (1992)
22. Wang, J., Li, H.: Solving simultaneous linear equations using recurrent neural networks. Inf. Sci. Intell. Syst. **76**(3), 255–277 (1994)
23. Xia, Y., Wang, J., Hung, D.L.: Recurrent neural networks for solving linear inequalities and equations. IEEE Trans. Circ. Syst. I Fundam. Theory Appl. **46**(4), 452–462 (1999)
24. Censor, Y., Elfving, T.: New methods for linear inequalities. Linear Algebra Appl. **42**, 199–211 (1982)
25. Anderson, E., Bai, Z., Bischof, C.: LAPACK Users' Guide. Software, environments, tools. Society for Industrial and Applied Mathematics, Philadelphia, PA (1999)
26. Lundberg, K.H.: The history of analog computing: introduction to the special section. IEEE Control Syst. **25**(3), 22–25 (June 2005)
27. Cowan, G.E.R., Melville, R.C., Tsividis, Y.P.: A VLSI analog computer/math co-processor for a digital computer. In: IEEE International Solid-State Circuits Conference (ISSCC), vol. 1, pp. 82–86 (2005)
28. Ansari, M.S., Rahman, S.A.: Integrable non-linear feedback analog circuit for solving linear equations. In: 2011 International Conference on Multimedia, Signal Processing and Communication Technologies (IMPACT), pp. 284–287 (2011)
29. Ansari, M.S., Rahman, S.A.: MO-OTA based recurrent neural network for solving simultaneous linear equations. In: 2011 International Conference on Multimedia, Signal Processing and Communication Technologies (IMPACT), pp. 192–195 (2011)
30. Newcomb, R.W., Lohn, J.D.: Analog VLSI for neural networks. In: Arbib, M.A. (ed.) The handbook of Brain Theory and Neural Networks, pp. 86–90. MIT Press, Cambridge (1998)

Chapter 4
Mixed-Mode Neural Circuit for Solving Linear Equations

4.1 Introduction

The voltage-mode neural network for solving linear equations presented in the previous chapter requires a number of passive resistors for the synaptic interconnections. This translates to larger chip area and component matching issues during the actual integrated circuit implementation of the circuit. Knowing that the synaptic resistances are used to generate currents corresponding to the output voltages of the comparators, such resistors may be eliminated by having a *mixed*-mode implementation wherein the neuron states are represented by node voltages and the synaptic signals are conveyed as currents. For that purpose, the voltage-mode comparators employed in the non-linear feedback connections could be replaced by voltage comparators with current outputs. An analog building block capable of such a functionality is the Differential Voltage Current Conveyor (DVCC) [1–5]. In this chapter, DVCC–based voltage comparators with current outputs are employed to obtain circuits with reduced complexity over their counterparts presented in Chap. 3.

This chapter is organized as follows. A mixed-mode implementation of the linear equation solver of Chap. 3, based on DVCC, is presented in Sect. 4.2. Results of PSPICE simulations carried out to ascertain the proper working of the circuit are presented in Sect. 4.3. To incorporate reconfigurability into the DVCC-based mixed-mode circuit for solving liner equations, a new digitally-programmable analog building block (DC-DVCC) is introduced in Sect. 4.4. Section 4.5 contains the details of the digitally programmable neural network for solving linear equations along with the design details. Proof of the energy function and validity of the solution are provided in the same section. Section 4.6 presents the results of PSPICE simulation of the circuit applied to solve various sample equation sets. A discussion on VLSI implementability of the circuit appears in Sect. 4.7. Some conclusive remarks appear in Sect. 4.8.

M. S. Ansari, *Non-Linear Feedback Neural Networks*,
Studies in Computational Intelligence 508, DOI: 10.1007/978-81-322-1563-9_4,
© Springer India 2014

Fig. 4.1 Symbolic
representation of DVCC

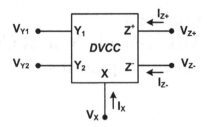

4.2 Mixed-Mode Neural Network for Solving Linear Equations

A voltage-mode neural network for solving systems of simultaneous linear equations, employing the NOSYNN architecture, was discussed in the previous chapter in which weighted synaptic resistors were employed to generate currents corresponding to the output voltages of the comparators. However, the requirement of a large number of resistances for the synaptic interconnections makes the circuit unattractive from the viewpoint of monolithic integration. The circuit complexity of the linear equation solver of Fig. 3.8 may further be reduced by replacing each voltage-mode comparator and the associated weight resistance by a transconductance element giving current output while accepting voltage inputs. The Differential Voltage Current Conveyor (DVCC) is one such analog building block [1–5]. The symbolic diagram of the DVCC is shown in Fig. 4.1 and its CMOS implementation is shown in Fig. 4.2 with the aspect ratios of the transistors given in Table 4.1. The device can be characterized by the following port relations:

$$
\begin{bmatrix} V_X \\ I_{Y1} \\ I_{Y2} \\ I_{Z+} \\ I_{Z-} \end{bmatrix} = \begin{bmatrix} 0 & 1 & -1 & 0 & 0 \\ 0 & 0 & 0 & 0 & 0 \\ 0 & 0 & 0 & 0 & 0 \\ 1 & 0 & 0 & 0 & 0 \\ -1 & 0 & 0 & 0 & 0 \end{bmatrix} \begin{bmatrix} I_X \\ V_{Y1} \\ V_{Y2} \\ V_{Z+} \\ V_{Z-} \end{bmatrix}
\tag{4.1}
$$

While the voltage at the X terminal is determined by the difference of the voltages at the Y_1 and Y_2 terminals, a current flowing in the X terminal is replicated at the $Z+$ terminal with the same directional polarity, and at $Z-$ terminal with a directional polarity opposite to that at the $Z+$ terminal. For instance, if the current I_X flows out of the X terminal, an equal valued current also flows out of the $Z+$ terminal, whereas an equal valued current flows into the $Z-$ terminal. Although both $Z+$ and $Z-$ types of current outputs are mentioned in (4.1), the DVCCs used in the network use only $Z+$ type of outputs.

DVCC as a voltage comparator with current outputs
For the DVCC, comparator action can be achieved by putting $R_X \rightarrow 0$ i.e. by directly grounding the X-terminal [2]. For such a case, the current in the X-port will saturate and can be written as

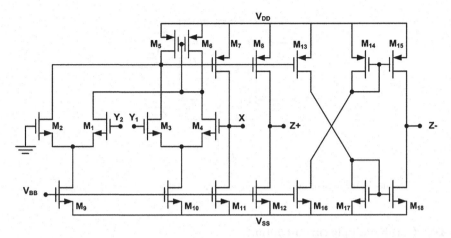

Fig. 4.2 CMOS implementation of DVCC

Table 4.1 Aspect ratios of the MOSFETs in the CMOS realization of the DVCC

Transistor	W(μm)/L(μm)
M1–M4	0.8/0.5
M5, M6	4/0.5
M9, M10	14.4/0.5
M7, M8, M13, M14, M15	10/0.5
M11, M12, M16, M17, M18	45/0.5

$$I_X = I_m \tanh \beta \ (V_{Y1} - V_{Y2}) \tag{4.2}$$

where β is the open-loop gain of the voltage comparator (practically very high) and $\pm I_m$ are the saturated output current levels of the DVCC. The value of $\pm I_m$ is governed by the bias current of the DVCC, which in turn is decided by the bias voltage V_{BB} in Fig. 4.2. Equation (4.2) can also be written in an equivalent notation as

$$I_X = g V_m \tanh \beta \ (V_{Y1} - V_{Y2}) \tag{4.3}$$

where g is the transconductance gain from the input voltage ports (Y_1 and Y_2) to the output current port (X) and $\pm V_m$ are the biasing voltages of the DVCC-based comparator. By virtue of DVCC action, this current will be transferred to the Z+ ports as

$$I_z^+ = I_X = g V_m \tanh \beta \ (V_{Y1} - V_{Y2}) \tag{4.4}$$

Therefore, the DVCC, with a grounded X terminal may be used as a voltage comparator with current output available at the Z+ terminal. The saturating values of the comparator characteristics are determined by the biasing supplies and the transconductance gain of the DVCC.

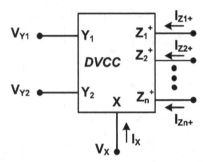

Fig. 4.3 Symbolic diagram of the DVCC with multiple Z+ outputs

DVCC with multiple current outputs

For use in the mixed-mode linear equation solver, multiple Z+ outputs are required. Figure 4.3 shows the symbolic diagram of such a DVCC with multiple Z+ outputs. The additional Z+ outputs can be readily obtained by replicating the output stage in the CMOS implementation of the DVCC which was presented in Fig. 4.2. To have the multiple Z+ outputs provide equal values currents, care must be exercised to keep the aspect ratios of the transistors in the replicated stages to be the same as in the original Z+ stage. The CMOS implementation of the DVCC with multiple Z+ outputs is presented in Fig. 4.4. A similar technique may be employed, and the Z− stage transistors replicated, if multiple Z− outputs are also desired.

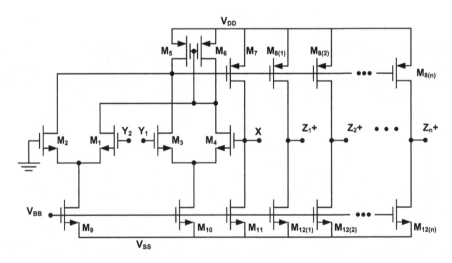

Fig. 4.4 CMOS implementation of the DVCC with multiple Z+ outputs

DVCC based mixed-mode circuit for solving linear equations
Let the simultaneous linear equations to be solved are

$$\mathbf{AV} = \mathbf{B} \tag{4.5}$$

where

$$\mathbf{A} = \begin{bmatrix} a_{11} & a_{12} & a_{13} & \cdots & a_{1n} \\ a_{21} & a_{22} & a_{23} & \cdots & a_{2n} \\ a_{31} & a_{32} & a_{33} & \cdots & a_{3n} \\ \vdots & \vdots & \vdots & \cdots & \vdots \\ a_{n1} & a_{n2} & a_{n3} & \cdots & a_{nn} \end{bmatrix} \tag{4.6}$$

$$\mathbf{B} = \begin{bmatrix} b_1 \\ b_2 \\ b_3 \\ \vdots \\ b_n \end{bmatrix} \tag{4.7}$$

$$\mathbf{V} = \begin{bmatrix} V_1 \\ V_2 \\ V_3 \\ \vdots \\ V_n \end{bmatrix} \tag{4.8}$$

where $V_1, V_2, \ldots, V_n$ are the decision variables and a_{ij} and b_i are constants. It may be mentioned that (4.5) is a compact representation of (2.82) and utilizes the matrix notation. Also, since the outputs of the neurons are voltages, the decision variables are now designated as voltages $V_1, V_2, \ldots, V_n$ to correspond to the output states of the neurons. It will be assumed that the coefficient matrix $\mathbf{A}$ is invertible, and hence, the system of linear Eq. (4.5) is consistent and not under-determined. In other words, the linear system (4.5) has a uniquely determined solution.

The mixed-mode neural circuit to solve the system of Equations of (4.5) is presented in Fig. 4.5 from where it can be seen that individual equations from the set of linear equations are passed through non-linear synapses which are realized using multi-output DVCC based bipolar voltage comparators. The outputs of the comparators are fed to neurons having weighted inputs. These weighted neurons are realized by using operaional amplifiers where the scaled currents coming from various comparators act as weights. R_{pi} and C_{pi} are the input resistance and capacitance of the opamp corresponding to the ith neuron. These parasitic components are included to model the dynamic nature of the opamp. As was done in the previous chapter too, all subsequent discussions regarding neuronal dynamics assume the fact that all the voltage comparators do not possess any delay and all the delay of the neuron is modelled at the neuron amplifier.

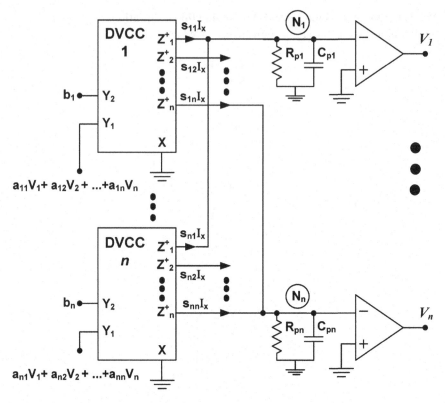

Fig. 4.5 DVCC-based mixed-mode feedback neural circuit to solve n simultaneous linear equations in n variables

Solution of two simultaneous linear equations in two variables
In order to understand the operation of the circuit, and also to obtain the energy function of the network, it is advisable to start from a simpler version of the general network shown in Fig. 4.5. For the sake of simplicity, we shall start with a circuit capable of solving two simultaneous linear equations in two variables. From Fig. 4.5, it is readily apparent that a circuit to solve a two variable problem would require 2 neuron amplifiers and 2 voltage comparators with current outputs. The inputs of the first comparator would be: a voltage equal to b_1 at the Y_2 terminal and $(a_{11}V_1 + a_{12}V_2)$ at the Y_1 terminal. The two current outputs of this first comparator are designated as $s_{11}I_x$ and $s_{12}I_x$. Similarly, for the second voltage-mode comparator, the inputs and outputs would be b_2 at the Y_2 terminal and $(a_{21}V_1 + a_{22}V_2)$ at the Y_1 terminal. The two current outputs of the second comparator will be $s_{21}I_x$ and $s_{22}I_x$. In due course, it shall be shown that the scaling factors for the output currents of the DVCC-based comparators, s_{11}, s_{12}, s_{21} and s_{22}, are determined by the coefficients a_{11}, a_{12}, a_{21} and a_{22}, in the system of linear equations to be solved.

The jth current output of the ith comparator in Fig. 4.5 can be modelled as

$$I_{ij} = s_{ij} \, g \, V_m \tanh \beta \, (a_{i1} V_1 + a_{i2} V_2 + \cdots + a_{in} V_n - b_i) \qquad (4.9)$$

where β is the open-loop gain of the comparator (practically very high), $\pm V_m$ are the output voltage levels of the comparator and $V_1, V_2, \ldots, V_n$ are the neuron outputs. Therefore, for a two variable linear equation solver, the two outputs of each of the two comparators would be given by

$$I_{11} = s_{11} \, g \, V_m \tanh \beta (a_{11} V_1 + a_{12} V_2 - b_1) \qquad (4.10)$$

$$I_{12} = s_{12} \, g \, V_m \tanh \beta (a_{11} V_1 + a_{12} V_2 - b_1) \qquad (4.11)$$

$$I_{21} = s_{21} \, g \, V_m \tanh \beta (a_{21} V_1 + a_{22} V_2 - b_1) \qquad (4.12)$$

$$I_{22} = s_{22} \, g \, V_m \tanh \beta (a_{21} V_1 + a_{22} V_2 - b_2) \qquad (4.13)$$

For the sake of comparison, let us reconsider the voltage-mode linear equation solver discussed in the previous chapter. The comparators in that circuit were voltage-mode comparators providing voltage outputs. Additional resistances were required to transform the comparator output voltages to weighted currents corresponding to the coefficients in the linear equations. The weighted currents were then added at the input of the neurons to provide steady-state voltages at the outputs of the neurons which were referred to as the solution points of the linear equations. In the present case of a DVCC-based mixed-mode circuit, it should be evident that the availability of currents at the outputs of the comparators alleviates the need of resistors connected in the synaptic interconnections between the neurons. This greatly lowers the overall complexity of the circuit due to reduction in the passive resistance count. The dynamics of this two neuron circuit can be understood by writing the equations of motion of the two neurons in state space.

At the input of the first neuron amplifier, we have

$$s_{11} I_X + s_{21} I_X = C_{p1} \frac{du_1}{dt} + \frac{u_1}{R_{p1}} \qquad (4.14)$$

which can be rearranged as

$$C_{p1} \frac{du_1}{dt} = s_{11} I_X + s_{21} I_X - \frac{u_1}{R_{p1}} \qquad (4.15)$$

Substituting the value of I_X found using (4.3) in (4.15), we get

$$C_{p1} \frac{du_1}{dt} = s_{11} \, g \, V_m \tanh \beta (a_{11} V_1 + a_{12} V_2 - b_1)$$

$$+ s_{21} \, g \, V_m \tanh \beta (a_{21} V_1 + a_{22} V_2 - b_2) - \frac{u_1}{R_{p1}} \qquad (4.16)$$

A similar analysis for the second neuron yields the following

$$C_{p2}\frac{du_2}{dt} = s_{12}\, g\, V_m \tanh \beta(a_{11}V_1 + a_{12}V_2 - b_2)$$

$$+ s_{22}\, g\, V_m \tanh \beta(a_{21}V_1 + a_{22}V_2 - b_2) - \frac{u_2}{R_{p2}} \qquad (4.17)$$

Furthermore, dynamical systems such as the one discussed here have an energy function which is related to the equation of motion of the two neurons as

$$\frac{\partial E}{\partial V_1} = KC_{p1}\frac{du_1}{dt}; \quad \frac{\partial E}{\partial V_2} = KC_{p2}\frac{du_2}{dt} \qquad (4.18)$$

where K is a multiplicative constant of proportionality with the dimensions of resistance [6, 7]. Following the approach of writing the energy function by inspection of the equation of motions of the neurons, one possible choice for the energy function could be

$$E_{2var} = \frac{V_m}{\beta} \ln \cosh \beta(a_{11}V_1 + a_{12}V_2 - b_1) + \frac{V_m}{\beta} \ln \cosh \beta(a_{21}V_1 + a_{22}V_2 - b_2)$$

$$- \frac{1}{R_{p1}} \int_0^{V_1} u_1 dV_1 - \frac{1}{R_{p2}} \int_0^{V_2} u_2 dV_2 \qquad (4.19)$$

The last two terms in (4.19) are only significant near the saturating values of the opamp and can otherwise be neglected for all operational voltages below the saturation voltage of the opamp [6, 7]. This results in a somewhat simpler expression of the energy function, as given in (4.20)

$$E_{2var,\,simplified} = \frac{V_m}{\beta} \ln \cosh \beta(a_{11}V_1 + a_{12}V_2 - b_1)$$

$$+ \frac{V_m}{\beta} \ln \cosh \beta(a_{21}V_1 + a_{22}V_2 - b_2) \qquad (4.20)$$

To verify whether the chosen energy function can indeed be associated with the two variable linear equation solver, we find $\partial E/\partial V_1$ and $\partial E/\partial V_2$ for the expression in (4.20) and then compare the result with the right-hand sides of (4.18), as shown below.

$$\frac{\partial E}{\partial V_1} = a_{11}V_m \tanh \beta(a_{11}V_1 + a_{12}V_2 - b_1) + a_{21}V_m \tanh \beta(a_{21}V_1 + a_{22}V_2 - b_2)$$

$$= KC_{p1}\frac{du_1}{dt} = Ks_{11}\, g\, V_m \tanh \beta(a_{11}V_1 + a_{12}V_2 - b_1)$$

$$+ Ks_{21}\, g\, V_m \tanh \beta(a_{21}V_1 + a_{22}V_2 - b_2) \qquad (4.21)$$

From (4.21), it can be observed that

$$a_{11} = K g s_{11}; \quad a_{21} = K g s_{21}$$

Therefore, the constant of proportionality K and the coefficients of the first variable V_1 in the set of linear equations to be solved, serve to set the values of the current scaling factors at the output of the first comparator. In fact, a suitable value of K may be chosen as

$$K = \frac{1}{g}$$

and then the values of the current scaling factors can be shown to be the same as the coefficients of the first variable in the system of linear equations.

A similar analysis is needed for the second neuron, wherein we start by finding $\partial E / \partial V_2$ and then equating it with the right-hand side of the second equality in (4.18).

$$\frac{\partial E}{\partial V_2} = a_{12} V_m \tanh \beta (a_{11} V_1 + a_{12} V_2 - b_1) + a_{22} V_m \tanh \beta (a_{21} V_1 + a_{22} V_2 - b_2)$$

$$= K C_{p2} \frac{du_2}{dt} = K g V_m \tanh \beta (a_{11} V_1 + a_{12} V_2 - b_1)$$

$$+ K g V_m \tanh \beta (a_{21} V_1 + a_{22} V_2 - b_2) \tag{4.22}$$

From (4.22), it can be observed that

$$a_{12} = K g s_{12}; \quad a_{22} = K g s_{22}$$

Therefore, for the second neuron too, the constant of proportionality K and the coefficients of the second variable V_2 in the set of linear equations to be solved, serve to set the actual values of the current scaling factor for the second DVCC based comparator. A suitable value of K may be chosen as $K = 1/g$ and the the values of s_{12} and s_{22} can shown to be equal to a_{12} and a_{22} respectively.

As an example, consider the following set of linear equations in 2 variables

$$\begin{bmatrix} 2 & 3 \\ 4 & 2 \end{bmatrix} \begin{bmatrix} V_1 \\ V_2 \end{bmatrix} = \begin{bmatrix} 7 \\ 12 \end{bmatrix} \tag{4.23}$$

In order to find the values of resistances that need to be used, we first let $K = 1/g$, where g is the transconductance gain associated with the DVCC and then compute the values of the current scaling parameters as

$$s_{11} = a_{11} = 2;$$
$$s_{12} = a_{12} = 3;$$
$$s_{21} = a_{21} = 4;$$
$$s_{22} = a_{22} = 2$$

The current scaling factors found above essentially imply that out of the two $Z+$ current outputs of the first DVCC-based comparator, the first $Z+$ output should have a current equal to 2 times the current flowing in the X terminal, and the second $Z+$ terminal should output a current which is 3 times (in magnitude) of the X terminal current. Similarly for the second DVCC-based comparator, the first and second $Z+$ terminals should output currents which are 4 and 2 times of the X terminal currents, respectively. Such a current scaling can be achieved by altering the aspect ratios of the transistors comprising the $Z+$ stages. For instance, to obtain a current scaling factor of 3 at the first $Z+$ terminal, the W/L ratios of the transistors $M_{8(1)}$ and $M_{12(1)}$ should be kept three times the aspect ratios of M_7 and M_{11} respectively.

It needs to be mentioned here that, like the case of the voltage-mode linear equations solver discussed in the previous chapter, some additional resistances would be required to generate the voltages, viz. $a_{11}V_1 + a_{12}V_2$ and $a_{21}V_1 + a_{22}V_2$, to be applied at the non-inverting inputs of the two DVCC based voltage comparators. As explained in the previous chapter, the required voltage summation may be achieved by using resistive potential dividers.

Solution of three linear equations in three variables
A 3 variable linear equation solver comprising of 3 DVCC-based comparators and 3 neuronal amplifiers is considered next. Each of the comparators are now required to have 3 scaled current outputs. The 3 current outputs of the first DVCC-based comparator may be written as

$$I_{11} = s_{11} \, g \, V_m \tanh \beta (a_{11}V_1 + a_{12}V_2 + a_{13}V_3 - b_1) \tag{4.24}$$

$$I_{12} = s_{12} \, g \, V_m \tanh \beta (a_{11}V_1 + a_{12}V_2 + a_{13}V_3 - b_1) \tag{4.25}$$

$$I_{13} = s_{13} \, g \, V_m \tanh \beta (a_{11}V_1 + a_{12}V_2 + a_{13}V_3 - b_1) \tag{4.26}$$

For the second DVCC based voltage comparator, the 3 current outputs are given as

$$I_{21} = s_{21} \, g \, V_m \tanh \beta (a_{21}V_1 + a_{22}V_2 + a_{23}V_3 - b_1) \tag{4.27}$$

$$I_{22} = s_{22} \, g \, V_m \tanh \beta (a_{21}V_1 + a_{22}V_2 + a_{23}V_3 - b_2) \tag{4.28}$$

$$I_{23} = s_{23} \, g \, V_m \tanh \beta (a_{21}V_1 + a_{22}V_2 + a_{23}V_3 - b_2) \tag{4.29}$$

Similarly, for the third comparator, the 3 current outputs are given by

$$I_{31} = s_{31} \, g \, V_m \tanh \beta (a_{31}V_1 + a_{32}V_2 + a_{33}V_3 - b_3) \tag{4.30}$$

$$I_{32} = s_{32} \, g \, V_m \tanh \beta (a_{31}V_1 + a_{32}V_2 + a_{33}V_3 - b_3) \tag{4.31}$$

$$I_{33} = s_{33} \, g \, V_m \tanh \beta (a_{31}V_1 + a_{32}V_2 + a_{33}V_3 - b_3) \tag{4.32}$$

As was done for the case of a 2 variable mixed-mode linear equation solver, equations of motion of the 3 neurons in the state space should be written to gain an insight into the dynamics of the network. At the input of the first neuron amplifier, we have

$$s_{11}I_X + s_{21}I_X + s_{31}I_X = C_{p1}\frac{du_1}{dt} + \frac{u_1}{R_{p1}} \tag{4.33}$$

which can be rearranged as

$$C_{p1}\frac{du_1}{dt} = s_{11}I_X + s_{21}I_X + s_{31}I_X - \frac{u_1}{R_{p1}} \tag{4.34}$$

Substituting the value of I_X found using (4.3) in (4.34), we get

$$C_{p1}\frac{du_1}{dt} = s_{11} g V_m \tanh \beta(a_{11}V_1 + a_{12}V_2 + a_{13}V_3 - b_1)$$

$$+ s_{21} g V_m \tanh \beta(a_{21}V_1 + a_{22}V_2 + a_{23}V_3 - b_2)$$

$$+ s_{31} g V_m \tanh \beta(a_{31}V_1 + a_{32}V_2 + a_{33}V_3 - b_3) - \frac{u_1}{R_{p1}} \tag{4.35}$$

A similar analysis for the second neuron yields the following

$$C_{p2}\frac{du_2}{dt} = s_{12} g V_m \tanh \beta(a_{11}V_1 + a_{12}V_2 + a_{13}V_3 - b_1)$$

$$+ s_{22} g V_m \tanh \beta(a_{21}V_1 + a_{22}V_2 + a_{23}V_3 - b_2)$$

$$+ s_{32} g V_m \tanh \beta(a_{31}V_1 + a_{32}V_2 + a_{33}V_3 - b_3) - \frac{u_2}{R_{p2}} \tag{4.36}$$

and lastly, for the third neuron, we have

$$C_{p3}\frac{du_3}{dt} = s_{13} g V_m \tanh \beta(a_{11}V_1 + a_{12}V_2 + a_{13}V_3 - b_1)$$

$$+ s_{23} g V_m \tanh \beta(a_{21}V_1 + a_{22}V_2 + a_{23}V_3 - b_2)$$

$$+ s_{33} g V_m \tanh \beta(a_{31}V_1 + a_{32}V_2 + a_{33}V_3 - b_3) - \frac{u_3}{R_{p3}} \tag{4.37}$$

Furthermore, as has already been mentioned, dynamical systems such as the one discussed here have an energy function which is related to the equation of motion of the three neurons as

$$\frac{\partial E}{\partial V_1} = KC_{p1}\frac{du_1}{dt}; \quad \frac{\partial E}{\partial V_2} = KC_{p2}\frac{du_2}{dt}; \quad \frac{\partial E}{\partial V_3} = KC_{p3}\frac{du_3}{dt}; \tag{4.38}$$

where K is a multiplicative constant of proportionality with the dimensions of resistance [6, 7]. Following the approach of writing the energy function by inspection of

the equation of motions of the neurons, one possible choice for the energy function
could be

$$
E_{3var} = \frac{V_m}{\beta} \ln \cosh \beta(a_{11}V_1 + a_{12}V_2 + a_{13}V_3 - b_1)
$$

$$
+ \frac{V_m}{\beta} \ln \cosh \beta(a_{21}V_1 + a_{22}V_2 + a_{23}V_3 - b_2)
$$

$$
+ \frac{V_m}{\beta} \ln \cosh \beta(a_{31}V_1 + a_{32}V_2 + a_{33}V_3 - b_3)
$$

$$
- \frac{1}{R_{p1}} \int_0^{V_1} u_1 dV_1 - \frac{1}{R_{p2}} \int_0^{V_2} u_2 dV_2 - \frac{1}{R_{p3}} \int_0^{V_3} u_3 dV_3 \qquad (4.39)
$$

The last three terms in (4.39) are only significant near the saturating values of
the opamp and can otherwise be neglected for all operational voltages below the
saturation voltage of the opamp [6, 7]. This results in a somewhat simpler expression
of the energy function, as given in (4.40)

$$
E_{3var,simplified} = \frac{V_m}{\beta} \ln \cosh \beta(a_{11}V_1 + a_{12}V_2 + a_{13}V_3 - b_1)
$$

$$
+ \frac{V_m}{\beta} \ln \cosh \beta(a_{21}V_1 + a_{22}V_2 + a_{23}V_3 - b_2)
$$

$$
+ \frac{V_m}{\beta} \ln \cosh \beta(a_{31}V_1 + a_{32}V_2 + a_{33}V_3 - b_3) \qquad (4.40)
$$

To verify whether the chosen energy function can indeed be associated with the
three variable linear equation solver, we find $\partial E/\partial V_1$, $\partial E/\partial V_2$ and $\partial E/\partial V_3$ for the
expression in (4.40) and then compare the result with the right-hand sides of (4.38),
as shown below.

$$
\frac{\partial E}{\partial V_1} = a_{11}V_m \tanh \beta(a_{11}V_1 + a_{12}V_2 + a_{13}V_3 - b_1)
$$

$$
+ a_{21}V_m \tanh \beta(a_{21}V_1 + a_{22}V_2 + a_{23}V_3 - b_2)
$$

$$
+ a_{31}V_m \tanh \beta(a_{31}V_1 + a_{32}V_2 + a_{33}V_3 - b_3)
$$

$$
= KC_{p1} \frac{du_1}{dt} = K s_{11} \, g \, V_m \tanh \beta(a_{11}V_1 + a_{12}V_2 + a_{13}V_3 - b_1)
$$

$$
+ K s_{21} \, g \, V_m \tanh \beta(a_{21}V_1 + a_{22}V_2 + a_{23}V_2 - b_3)
$$

$$
+ K s_{31} \, g \, V_m \tanh \beta(a_{31}V_1 + a_{32}V_2 + a_{33}V_3 - b_3) \qquad (4.41)
$$

From (4.41), it can be observed that

$$
a_{11} = K \, g \, s_{11}; \quad a_{21} = K \, g \, s_{21}; \quad a_{31} = K \, g \, s_{31}
$$

Therefore, the constant of proportionality K and the coefficients of the first variable V_1 in the set of linear equations to be solved, serve to set the values of the current scaling factors at the output of the first comparator. If the value of K is selected as $K = 1/g$, then the values of the current scaling factors can be shown to be the same as the coefficients of the first variable in the system of linear equations.

$$s_{11} = a_{11}; \quad s_{21} = a_{21}; \quad s_{31} = a_{31}$$

A similar analysis is needed for the second neuron, wherein we start by finding $\partial E / \partial V_2$ and then equating it with the right-hand side of the second equality in (4.18).

$$
\begin{aligned}
\frac{\partial E}{\partial V_2} &= a_{12} V_m \tanh \beta (a_{11} V_1 + a_{12} V_2 + a_{13} V_3 - b_1) \\
&\quad + a_{22} V_m \tanh \beta (a_{21} V_1 + a_{22} V_2 + a_{23} V_3 - b_2) \\
&\quad + a_{32} V_m \tanh \beta (a_{31} V_1 + a_{32} V_2 + a_{33} V_3 - b_3) \\
&= K C_{p2} \frac{du_2}{dt} = K s_{12} \ g \ V_m \tanh \beta (a_{11} V_1 + a_{12} V_2 + a_{13} V_3 - b_1) \\
&\quad + K s_{22} \ g \ V_m \tanh \beta (a_{21} V_1 + a_{22} V_2 + a_{23} V_2 - b_2) \\
&\quad + K s_{32} \ g \ V_m \tanh \beta (a_{31} V_1 + a_{32} V_2 + a_{33} V_3 - b_3) \quad (4.42)
\end{aligned}
$$

From (4.42), it can be observed that

$$a_{12} = K \ g \ s_{12}; \quad a_{22} = K \ g \ s_{22}; \quad a_{32} = K \ g \ s_{32}$$

For $K = 1/g$, we get

$$s_{12} = a_{12}; \quad s_{22} = a_{22}; \quad s_{32} = a_{32}$$

Lastly for the third neuron, we have

$$
\begin{aligned}
\frac{\partial E}{\partial V_3} &= a_{13} V_m \tanh \beta (a_{11} V_1 + a_{12} V_2 + a_{13} V_3 - b_1) \\
&\quad + a_{23} V_m \tanh \beta (a_{21} V_1 + a_{22} V_2 + a_{23} V_3 - b_2) \\
&\quad + a_{33} V_m \tanh \beta (a_{31} V_1 + a_{32} V_2 + a_{33} V_3 - b_3) \\
&= K C_{p3} \frac{du_3}{dt} = K s_{13} \ g \ V_m \tanh \beta (a_{11} V_1 + a_{12} V_2 + a_{13} V_3 - b_1) \\
&\quad + K s_{23} \ g \ V_m \tanh \beta (a_{21} V_1 + a_{22} V_2 + a_{23} V_2 - b_2) \\
&\quad + K s_{33} \ g \ V_m \tanh \beta (a_{31} V_1 + a_{32} V_2 + a_{33} V_3 - b_3) \quad (4.43)
\end{aligned}
$$

From (4.43), it can be observed that

$$a_{13} = K \ g \ s_{13}; \quad a_{23} = K \ g \ s_{23}; \quad a_{33} = K \ g \ s_{33}$$

For $K = 1/g$, we get

$$s_{13} = a_{13}; \quad s_{23} = a_{23}; \quad s_{33} = a_{33}$$

As an example, consider the following set of linear equations in 3 variables

$$\begin{bmatrix} 2 & 3 & 3 \\ 4 & 2 & 1 \\ 1 & 4 & 1 \end{bmatrix} \begin{bmatrix} V_1 \\ V_2 \\ V_3 \end{bmatrix} = \begin{bmatrix} 7 \\ 10 \\ 6 \end{bmatrix} \qquad (4.44)$$

In order to find the values of the current scaling factors for the 3 $Z+$ outputs of the thee comparators, we first let $K = 1/g$, where g is the transconductance gain associated with the DVCC and then compute the values of the current scaling parameters as

$$s_{11} = a_{11} = 2;$$
$$s_{12} = a_{12} = 3;$$
$$s_{13} = a_{13} = 3;$$
$$s_{21} = a_{21} = 4;$$
$$s_{22} = a_{22} = 2;$$
$$s_{23} = a_{23} = 1;$$
$$s_{31} = a_{31} = 1;$$
$$s_{32} = a_{32} = 4;$$
$$s_{33} = a_{33} = 1$$

The current scaling factors found above essentially imply that out of the three $Z+$ current outputs of the first DVCC-based comparator, the first $Z+$ output should have a current equal to 2 times the current flowing in the X terminal, the second $Z+$ terminal should output a current which is 3 times (in magnitude) of the X terminal current, and the third current output should be 3 times the X terminal current. As mentioned earlier, the different values of current scaling factors can be achieved by altering the aspect ratios of the transistors comprising the $Z+$ stages.

General case: n equations in n variables

Having explained the energy function and the method to obtain the values of the current scaling factors to be set for the DVCC-based comparators in the circuits to solve 2 and 3 variable systems of equations, we now move on to a generic network which can be used to solve n linear equations by employing n neurons connected through n non-linear synapses. The circuit diagram for the general network has already been presented in Fig. 4.5 from where it can be seen that n operational amplifiers are required to emulate the functionality of n neurons and n voltage comparators with multiple, scalable, current outputs (implemented using multiple-output DVCCs) are needed to provide the required non-linear synaptic feedback from the output of the neurons to the inputs of other neurons.

To find the values of the various scaling factors (s_{ij}) in Fig. 4.5, we follow a procedure similar to that employed in the case of 2 and 3 variable linear equation solvers discussed previously, and start by writing the equations of motion of the ith neuron in the state space and then associate a valid energy function with the network. Once an energy function is ascertained, the values of the current scaling factors are then determined by correlating the terms in the partial differential of the energy function with respect to each decision variable, with their counterparts in the equation of motion for each neuron.

Applying KCL at the inverting input terminal of the ith operational amplifier in Fig. 4.5, results in the following equation of motion of the ith neuron in the state space:

$$C_{pi}\frac{du_i}{dt} = s_{1i}I_{X1} + s_{2i}I_{X2} + \cdots + s_{ni}I_{Xn} - \left[\frac{u_i}{R_{pi}}\right] \tag{4.45}$$

where u_i is the internal state of the ith neuron and s_{ji} is the current scaling factor at the jth output of ith DVCC. The required current scaling at a particular Z+ ouput is obtained by altering the aspect ratios of the MOSFETs in the corresponding output stage in the CMOS implementation of Fig. 4.4. Using (4.3) in (4.45) results in

$$C_{pi}\frac{du_i}{dt} = s_{1i}\,gV_m\tanh\beta(a_{11}V_1 + a_{12}V_2 + \cdots + a_{1n}V_n - b_1)$$

$$+ s_{2i}\,gV_m\tanh\beta(a_{21}V_1 + a_{22}V_2 + \cdots + a_{2n}V_n - b_2) + \cdots +$$

$$+ s_{ni}\,gV_m\tanh\beta\,(a_{n1}V_1 + a_{n2}V_2 + \cdots + a_{nn}V_n - b_n) - \frac{u_i}{R_{pi}} \tag{4.46}$$

Starting from (4.46), a suitable energy function for the network in Fig. 4.5 can be written as

$$E = \frac{V_m}{\beta}\sum_{i=1}^{n}\ln\cosh\beta\left(\sum_{=1}^{n}a_{ij}V_j - b_i\right) - \sum_{i=1}^{n}\frac{1}{R_i}\int_0^{V_i}u_i dV_i \tag{4.47}$$

From (4.47), it follows that

$$\frac{\partial E}{\partial V_i} = V_m a_{1i}\tanh\beta(a_{11}V_1 + a_{12}V_2 + \cdots + a_{1n}V_n - b_1) +$$

$$V_m a_{2i}\tanh\beta(a_{21}V_1 + a_{22}V_2 + \cdots + a_{2n}V_n - b_2) + \cdots +$$

$$V_m a_{ni}\tanh\beta(a_{n1}V_1 + a_{n2}V_2 + \cdots + a_{nn}V_n - b_n) - \frac{u_i}{R_{pi}} \tag{4.48}$$

Also, if E is the Energy Function, it must satisfy the following condition [6].

$$\frac{\partial E}{\partial V_i} = KC_{pi}\frac{du_i}{dt} \tag{4.49}$$

where K is a constant of proportionality having the dimensions of resistance and is normalized to $1/g$ for simplicity. Comparing (4.46) and (4.48) according to (4.49) yields

$$\begin{bmatrix} s_{11} & s_{12} & \cdots & s_{1n} \\ s_{21} & s_{22} & \cdots & s_{2n} \\ \vdots & \vdots & \cdots & \vdots \\ s_{n1} & s_{n2} & \cdots & s_{nn} \end{bmatrix} = \begin{bmatrix} a_{11} & a_{12} & \cdots & a_{1n} \\ a_{21} & a_{22} & \cdots & a_{2n} \\ \vdots & \vdots & \cdots & \vdots \\ a_{n1} & a_{n2} & \cdots & a_{nn} \end{bmatrix} \tag{4.50}$$

4.2.1 Proof of the Energy Function

For E to be a valid energy function, it must satisfy two criteria. Firstly, it should be bounded from below, and secondly, it should be non-increasing with time. The first criterion viz. the energy function must have a lower bound is also satisfied for the circuit of Fig. 4.5 wherein it may be seen that $V_1, V_2, \ldots, V_n$ are all bounded (as they are the outputs of operational amplifiers) amounting to E, as given in (4.47), having a defined lower bound, which is decided by the biasing supplies of the opamp.

To verify that the chosen energy function E satisfies the 'non-increasing-with-time' criterion, we proceed by finding the time derivative of the energy function, as given below.

$$\frac{dE}{dt} = \sum_{i=1}^{N} \frac{\partial E}{\partial V_i} \frac{dV_i}{dt} = \sum_{i=1}^{N} \frac{\partial E}{\partial V_i} \frac{dV_i}{du_i} \frac{du_i}{dt} \tag{4.51}$$

Using (4.49) in (4.51) we get

$$\frac{dE}{dt} = \sum_{i=1}^{N} K \, C_{pi} \left(\frac{du_i}{dt} \right)^2 \frac{dV_i}{du_i} \tag{4.52}$$

The transfer characteristics of the output opamp used in Fig. 4.5 implements the activation function of the neuron. With u_i being the inverting terminal, it is monotonically decreasing and it can be seen that [6, 7],

$$\frac{dV_i}{du_i} \leq 0 \tag{4.53}$$

thereby resulting in

$$\frac{dE}{dt} \leq 0 \tag{4.54}$$

with the equality being valid for

$$\frac{du_i}{dt} = 0; \quad \text{for all } i \tag{4.55}$$

Equation (4.54) shows that the energy function can never increase with time which is one of the conditions for a valid energy function. Therefore, we may say with certainty that the Lyapunov function E is indeed a valid energy function. The fact that E is associated with the circuit of Fig. 4.5 has already been discussed in the previous section where it was shown that the partial derivative of E with respect to the decision variables is the same as the equation of motion of the individual neurons.

4.2.2 Stable States of the Network

Convergence of the network to the global minimum of the Energy Function, which is exactly the solution of the set of linear equations, and the fact that there are no other minima, can be shown as follows.

Simple Case: 2 linear equations in 2 variables

It has already been shown in (4.20) that the significant terms in the energy function for the case of a 2 variable linear equation solver are given by

$$E_{2var,simplified} = \frac{V_m}{\beta} \ln \cosh \beta (a_{11} V_1 + a_{12} V_2 - b_1)$$

$$+ \frac{V_m}{\beta} \ln \cosh \beta (a_{21} V_1 + a_{22} V_2 - b_2) \qquad (4.56)$$

from which, it follows that

$$\frac{\partial E}{\partial V_1} = a_{11} V_m \tanh \beta (a_{11} V_1 + a_{12} V_2 - b_1)$$

$$+ a_{21} V_m \tanh \beta (a_{21} V_1 + a_{22} V_2 - b_2) \qquad (4.57)$$

and

$$\frac{\partial E}{\partial V_2} = a_{12} V_m \tanh \beta (a_{11} V_1 + a_{12} V_2 - b_1)$$

$$+ a_{22} V_m \tanh \beta (a_{21} V_1 + a_{22} V_2 - b_2) \qquad (4.58)$$

For a stationary point, the partial derivatives of E with respect to both the decision variables must be zero.

$$\frac{\partial E}{\partial V_1} = 0; \quad \frac{\partial E}{\partial V_2} = 0; \qquad (4.59)$$

Using (4.57) and (4.58) in (4.59), we get

$$a_{11} V_m \tanh \beta (a_{11} V_1 + a_{12} V_2 - b_1) + a_{21} V_m \tanh \beta (a_{21} V_1 + a_{22} V_2 - b_2) = 0$$
$$a_{12} V_m \tanh \beta (a_{11} V_1 + a_{12} V_2 - b_1) + a_{22} V_m \tanh \beta (a_{21} V_1 + a_{22} V_2 - b_2) = 0$$
$$\text{(4.60)}$$

For the sake of simplicity, we denote $V_m \tanh \beta (a_{11} V_1 + a_{12} V_2 - b_1)$ by A_1 and $V_m \tanh \beta (a_{21} V_1 + a_{22} V_2 - b_2)$ by A_2, and thus (4.60) can be simplified to

$$a_{11} A_1 + a_{21} A_2 = 0$$
$$a_{12} A_1 + a_{22} A_2 = 0 \qquad \text{(4.61)}$$

Equation (4.61) may be represented in matrix notation as

$$\begin{bmatrix} a_{11} & a_{21} \\ a_{12} & a_{22} \end{bmatrix} \begin{bmatrix} A_1 \\ A_2 \end{bmatrix} = \begin{bmatrix} 0 \\ 0 \end{bmatrix} \qquad \text{(4.62)}$$

Since the above is a homogeneous system of linear equations, we can easily obtain the trivial solution which is

$$\begin{bmatrix} A_1 \\ A_2 \end{bmatrix} = \begin{bmatrix} 0 \\ 0 \end{bmatrix} \qquad \text{(4.63)}$$

which can be expanded to

$$V_m \tanh \beta (a_{11} V_1 + a_{12} V_2 - b_1) = 0$$
$$V_m \tanh \beta (a_{12} V_1 + a_{22} V_2 - b_2) = 0 \qquad \text{(4.64)}$$

and then simplified to
$$a_{11} V_1 + a_{12} V_2 - b_1 = 0$$
$$a_{12} V_1 + a_{22} V_2 - b_2 = 0 \qquad \text{(4.65)}$$

It is readily verified that the stationary point obtained in (4.65) is the same as the solution of the system of equations in 2 variables. Next, a similar proof is presented for the generalized case of a NOSYNN based neural network comprising on n neurons and n synapses, employed to solve a system of n simultaneous linear equations in n variables.

General Case: n linear equations in n variables
For a NOSYNN based neural circuit for solving n linear equations in n variables, the energy function is given by (4.47). However, the second term in (4.47) is significant only near the saturating values of the opamp and is usually neglected [8]. The energy function can therefore be expressed as

$$E = \frac{V_m}{\beta} \sum_{i=1}^{n} \ln \cosh \beta \left(\sum_{j=1}^{n} a_{ij} V_j - b_i \right) \qquad \text{(4.66)}$$

From which it follows that

$$\frac{\partial E}{\partial V_i} = \frac{V_m}{\beta} \, a_{1i} \, \tanh \beta(a_{11}V_1 + a_{12}V_2 + \cdots + a_{1n}V_n - b_1)$$

$$+ \frac{V_m}{\beta} \, a_{2i} \, \tanh \beta(a_{21}V_1 + a_{22}V_2 + \cdots + a_{2n}V_n - b_2) + \cdots$$

$$+ \frac{V_m}{\beta} \, a_{ni} \, \tanh \beta(a_{n1}V_1 + a_{n2}V_2 + \cdots + a_{nn}V_n - b_n) \quad (4.67)$$

For a stationary point, we have

$$\frac{\partial E}{\partial V_i} = 0 \quad (4.68)$$

which yields,

$$a_{1i} \, \tanh \beta(a_{11}V_1 + a_{12}V_2 + \cdots + a_{1n}V_n - b_1)$$
$$+ \, a_{2i} \, \tanh \beta(a_{21}V_1 + a_{22}V_2 + \cdots + a_{2n}V_n - b_2) + \cdots$$
$$+ \, a_{ni} \, \tanh \beta(a_{n1}V_1 + a_{n2}V_2 + \cdots + a_{nn}V_n - b_n) = 0 \quad (4.69)$$

Denoting

$$\tanh \beta(a_{11} V_1 + a_{12} V_2 + \cdots + a_{1n}V_n - b_1) = A_1$$
$$\tanh \beta(a_{21} V_1 + a_{22} V_2 + \cdots + a_{2n}V_n - b_2) = A_2$$
$$\vdots$$
$$\tanh \beta(a_{n1} V_1 + a_{n2} V_2 + \cdots + a_{nn}V_n - b_n) = A_n \quad (4.70)$$

Therefore, for a stationary point we have,

$$\begin{bmatrix} a_{11} \, a_{12} \, \ldots \, a_{1n} \\ a_{21} \, a_{22} \, \ldots \, a_{2n} \\ \vdots \quad \vdots \quad \ldots \quad \vdots \\ a_{n1} \, a_{n2} \, \ldots \, a_{nn} \end{bmatrix} \begin{bmatrix} A_1 \\ A_2 \\ \vdots \\ A_n \end{bmatrix} = \begin{bmatrix} 0 \\ 0 \\ \vdots \\ 0 \end{bmatrix} \quad (4.71)$$

This is a homogeneous system of linear equations in variables $A_1, A_2, \ldots, A_n$. Since the coefficient matrix of the set of equations (4.71) is the same as that of (4.5) which is invertible, it follows that (4.71) will have a uniquely determined solution which is the trivial solution of the homogeneous system. Therefore,

$$\begin{bmatrix} A_1 \\ A_2 \\ \vdots \\ A_n \end{bmatrix} = \begin{bmatrix} 0 \\ 0 \\ \vdots \\ 0 \end{bmatrix} \quad (4.72)$$

which results in,

$$a_{11}V_1 + a_{12}V_2 + \cdots + a_{1n}V_n - b_1 = 0$$
$$a_{21}V_1 + a_{22}V_2 + \cdots + a_{2n}V_n - b_2 = 0$$
$$\vdots$$
$$a_{n1}V_1 + a_{n2}V_2 + \cdots + a_{nn}V_n - b_n = 0 \qquad (4.73)$$

Thus, the energy function of the neural network has a unique stationary point which coincides exactly with the solution of the given system of linear equations. The fact that there is only a single minimum in the energy function makes this network significantly better than Hopfield Neural Network based approaches where the problem of multiple local minima is common. Also, the presence of transcendental terms in the energy function of (4.66), as opposed to quadratic terms in the energy function associated with the standard Hopfield Network, is responsible for better convergence characteristics in the NOSYNN based network.

4.3 Hardware Simulation Results

The operation of the network of Fig. 4.5 was verified through computer simulations using PSPICE program. For the purpose of PSPICE simulations, use was made of the LMC7101A CMOS opamp from National Semiconductor [9]. The sub-circuit file for this opamp is available in Orcad Model Library. For the multi-output DVCC of Fig. 4.4, standard 0.5 μm CMOS parameters were used for simulation purposes and the value of g was measured to be 1.092 milli-mhos. For the purpose of simulations, various sets of simultaneous linear equations in 2 to 10 variables were selected. By altering the aspect ratios of the transistors used in the i-th output stage of the j-th DVCC, I_{Zji} can be made a scaled replica of I_{Xj} . In that case, the value of s_{ji} will be different from unity and can be set to the required value according to (4.50).

PSPICE Simulation Result: 2 linear equations in 2 variables
A simple circuit comprising of 2 neurons (realized using operational amplifiers) and 2 non-linear synapses (comprising of voltage comparators with multiple $Z+$ outputs) was first tested using computer simulations using the PSPICE program. Consider the following system of linear equations in 2 variables:

$$\begin{bmatrix} 1 & 2 \\ 2 & 1 \end{bmatrix} \begin{bmatrix} V_1 \\ V_2 \end{bmatrix} = \begin{bmatrix} 3.5 \\ 4 \end{bmatrix} \qquad (4.74)$$

The mathematical solution of (4.74) is $V_1 = +1.5$ and $V_2 = +1$. The values of the current scaling factors required to be set for the two outputs of the two comparators can be found as.

$$s_{11} = a_{11} = 1;\ s_{12} = a_{12} = 2;\ s_{21} = a_{21} = 2;\ s_{22} = a_{22} = 1$$

The scaling factors given above may be implemented in the actual CMOS realization by proper setting of the aspect ratios of the transistors employed in the $Z+$ stages of the DVCC based comparators. For instance, for the first DVCC, the second $Z+$ stage transistors should have an area double that of the X stage transistors, whereas the first $Z+$ stage transistors would have W/L ratios equal to their counterparts in the X stage.

As has already been explained in the previous chapter, some additional circuitry is needed to generate the voltage required at the input of the comparators. This can be done by a resistive potential divider, as is explained next. We shall consider a nomenclature similar to that used in the case of the voltage-mode linear equation solver. To obtain the values of the resistances required at the input of the first comparator, we proceed by properly scaling the first equation from the set of linear equations (4.74). Selecting a scaling factor of 5, we have

$$\frac{1}{5}V_1 + \frac{2}{5}V_2 = \frac{3.5}{5} \tag{4.75}$$

from where the values of the resistances to be used in the potential divider can be obtained as

$$R_{e11} = 5\,K\Omega$$
$$R_{e12} = 2.5\,K\Omega$$
$$R_{e13} = 2.5\,K\Omega$$

and the voltage to be applied at the Y_2 input terminal of the first DVCC based comparator would be

$$b_1 = \frac{3.5}{5} = 0.7\,V$$

Similarly, scaling the second equation by a factor of 5, we have

$$\frac{2}{5}V_1 + \frac{1}{5}V_2 = \frac{4}{5} \tag{4.76}$$

from where the values of the resistances to be used in the potential divider can be obtained as

$$R_{e21} = 2.5\,K\Omega$$
$$R_{e22} = 5\,K\Omega$$
$$R_{e23} = 2.5\,K\Omega$$

and the voltage to be applied at the Y_2 input terminal of the second DVCC based comparator would be

$$b_2 = \frac{4}{5} = 0.8 \, V$$

The mixed-mode circuit was tested in PSPICE and the steady state neuronal output voltages were obtained as $V_1 = 1.50$ and $V_2 = 1.02 \, V$, which are in close agreement with the mathematical results for the chosen set of linear equations.

PSPICE Simulation Result: 3 linear equations in 3 variables
A simple circuit comprising of 3 neurons (realized using operational amplifiers) and 3 non-linear synapses (comprising of voltage comparators with multiple $Z+$ outputs) was then tested using PSPICE. Consider the following system of linear equations in 3 variables:

$$\begin{bmatrix} 2 & 1 & 1 \\ 3 & 2 & 1 \\ 1 & 1 & 2 \end{bmatrix} \begin{bmatrix} V_1 \\ V_2 \\ V_3 \end{bmatrix} = \begin{bmatrix} 5 \\ 10 \\ 6 \end{bmatrix} \tag{4.77}$$

The mathematical solution of (4.77) is $V_1 = -0.5$, $V_1 = +5.5$ and $V_3 = +0.5$. The values of the current scaling factors required to be set for the three outputs of the first DVCC-based comparator can be found as.

$$s_{11} = a_{11} = 2$$
$$s_{12} = a_{12} = 1$$
$$s_{13} = a_{13} = 1$$

and for the second comparator, the three $Z+$ outputs need to have current scaling factors as given below.

$$s_{21} = a_{21} = 3$$
$$s_{22} = a_{22} = 2$$
$$s_{23} = a_{23} = 1$$

Lastly, for the third comparator, the three current scaling factors are given by

$$s_{31} = a_{31} = 1$$
$$s_{32} = a_{32} = 1$$
$$s_{33} = a_{33} = 2$$

As discussed earlier, these scaling factors may be implemented in the actual CMOS realization by proper setting of the aspect ratios of the transistors employed in the $Z+$ stages of the DVCC based comparators. Further, to obtain the values of the resistances required at the input of the first comparator, we proceed by properly scaling the first equation from the set of linear Eq. (4.77). Selecting a scaling factor of 5, we have

$$\frac{2}{5}V_1 + \frac{1}{5}V_2 + \frac{1}{5}V_3 = \frac{5}{5} \tag{4.78}$$

from where the values of the resistances to be used in the potential divider can be obtained as

$$R_{e11} = 2.5 \, K\Omega$$
$$R_{e12} = 5 \, K\Omega$$
$$R_{e13} = 5 \, K\Omega$$
$$R_{e14} = 5 \, K\Omega$$

and the voltage to be applied at the Y_2 input terminal of the first DVCC based comparator would be

$$b_1 = \frac{5}{5} = 1 \, V$$

Similarly, scaling the second equation by a factor of 6, we have

$$\frac{3}{6}V_1 + \frac{2}{6}V_2 + \frac{1}{6}V_3 = \frac{10}{6} \tag{4.79}$$

from where the values of the resistances to be used in the potential divider can be obtained as

$$R_{e21} = 2 \, K\Omega$$
$$R_{e22} = 3 \, K\Omega$$
$$R_{e23} = 6 \, K\Omega$$
$$R_{e24} = \infty$$

and the voltage to be applied at the Y_2 input terminal of the second DVCC based comparator would be

$$b_2 = \frac{10}{6} = 1.66 \, V$$

Lastly, scaling the third equation by a factor of 6, we have

$$\frac{1}{6}V_1 + \frac{1}{6}V_2 + \frac{2}{6}V_3 = \frac{6}{6} \tag{4.80}$$

from where the values of the resistances to be used in the potential divider can be obtained as

$$R_{e21} = 6 \, K\Omega$$
$$R_{e22} = 6 \, K\Omega$$
$$R_{e23} = 3 \, K\Omega$$
$$R_{e24} = 3 \, K\Omega$$

and the voltage to be applied at the Y_2 input terminal of the third DVCC based comparator would be

$$b_3 = \frac{6}{6} = 1 \ V$$

The mixed-mode circuit was tested in PSPICE and the steady state neuronal output voltages were obtained as $V_1 = -0.51 \ V$, $V_2 = 5.53 \ V$ and $V_3 = 0.49 \ V$, which are in close proximity with the mathematical results for the 3 variable system of linear equations.

Results of simulation runs for these problems, as well as some other problems in higher variable counts, are presented in Table 4.2. Each of the simulations was run using various initial conditions in the millivolt range. As can be seen from Table 4.2, the network always converges to the solution of the given system of linear equations. Also, it can be seen that, the obtained results are in good agreement with the algebraic solutions in all the test cases.

Table 4.2 PSPICE simulation results for the DVCC based mixed-mode linear equation solver circuit applied to solve different systems of linear equations

[A]	[B]	Algebraic solution [V]	Simulated results (using PSPICE) [V]	Percentage error in the solution (%)
$\begin{bmatrix} 1 & 2 \\ 2 & 1 \end{bmatrix}$	$\begin{bmatrix} 3.5 \\ 4 \end{bmatrix}$	$\begin{bmatrix} 1.5 \\ 1 \end{bmatrix}$	$\begin{bmatrix} 1.50 \\ 1.02 \end{bmatrix}$	$\begin{bmatrix} 0 \\ 2.00 \end{bmatrix}$
$\begin{bmatrix} 2 & 1 & 1 \\ 3 & 2 & 1 \\ 1 & 1 & 2 \end{bmatrix}$	$\begin{bmatrix} 5 \\ 10 \\ 6 \end{bmatrix}$	$\begin{bmatrix} -0.5 \\ 5.5 \\ 0.5 \end{bmatrix}$	$\begin{bmatrix} -0.51 \\ 5.53 \\ 0.49 \end{bmatrix}$	$\begin{bmatrix} 2 \\ 0.54 \\ -2 \end{bmatrix}$
$\begin{bmatrix} 2 & 3 & 6 & 1 \\ 3 & 2 & 5 & 4 \\ 2 & 5 & 5 & 2 \\ 2 & 4 & 2 & 5 \end{bmatrix}$	$\begin{bmatrix} 48.75 \\ 67.5 \\ 54 \\ 59.25 \end{bmatrix}$	$\begin{bmatrix} 3.5 \\ 1.5 \\ 5 \\ 7.25 \end{bmatrix}$	$\begin{bmatrix} 3.52 \\ 1.51 \\ 4.98 \\ 7.27 \end{bmatrix}$	$\begin{bmatrix} 0.57 \\ 0.67 \\ -0.40 \\ 0.27 \end{bmatrix}$
$\begin{bmatrix} 2 & 3 & 9 & 2 & 5 \\ 2 & 6 & 9 & 9 & 5 \\ 2 & 6 & 2 & 4 & 5 \\ 2 & 4 & 7 & 8 & 3 \\ 5 & 3 & 6 & 3 & 5 \end{bmatrix}$	$\begin{bmatrix} -54.9 \\ -94.0 \\ -54.9 \\ -74.8 \\ -70.0 \end{bmatrix}$	$\begin{bmatrix} -6.7 \\ -4.3 \\ -2.8 \\ -3.3 \\ 0.64 \end{bmatrix}$	$\begin{bmatrix} -6.54 \\ -4.23 \\ -2.86 \\ -3.35 \\ 0.62 \end{bmatrix}$	$\begin{bmatrix} -2.38 \\ -1.62 \\ 2.14 \\ 1.51 \\ -3.12 \end{bmatrix}$
$\begin{bmatrix} 1 & 2 & 1 & 3 & 4 & 2 & 1 & 1 & 1 & 2 \\ 2 & 1 & 1 & 3 & 1 & 2 & 2 & 1 & 2 & 3 \\ 1 & 1 & 4 & 3 & 3 & 1 & 1 & 4 & 4 & 1 \\ 4 & 2 & 1 & 5 & 3 & 3 & 1 & 1 & 2 & 2 \\ 1 & 1 & 5 & 1 & 2 & 1 & 2 & 2 & 5 & 2 \\ 3 & 3 & 1 & 2 & 1 & 1 & 5 & 2 & 1 & 1 \\ 5 & 5 & 1 & 4 & 1 & 1 & 3 & 4 & 2 & 2 \\ 1 & 1 & 1 & 2 & 2 & 2 & 3 & 3 & 4 & 1 \\ 4 & 2 & 2 & 1 & 3 & 2 & 5 & 4 & 3 & 2 \\ 1 & 2 & 3 & 1 & 2 & 3 & 1 & 2 & 3 & 4 \end{bmatrix}$	$\begin{bmatrix} 10 \\ -11 \\ 10 \\ 2 \\ -7 \\ -9 \\ -8 \\ -3 \\ -3 \\ 7 \end{bmatrix}$	$\begin{bmatrix} -2 \\ 1 \\ 3 \\ -1 \\ 2 \\ 7 \\ -3 \\ 4 \\ -5 \\ -4 \end{bmatrix}$	$\begin{bmatrix} -1.92 \\ 1.02 \\ 3.11 \\ -1.01 \\ 2.02 \\ 7.05 \\ -3.08 \\ 4.12 \\ -4.91 \\ -3.94 \end{bmatrix}$	$\begin{bmatrix} -4 \\ 2 \\ 3.66 \\ 1 \\ 1 \\ 0.71 \\ 2.66 \\ 3 \\ -1.80 \\ -1.50 \end{bmatrix}$

At this juncture, it is important to point out the technique of altering the aspect ratios of the transistors to obtain the required amount of current scaling in only a theoretical concept with little practical significance. This is because, the circuit once fabricated for a particular set of scaling factors, can not be reconfigured for any other ones simply because of the fact that the transistors areas could not be changed once fabricated. Therefore, in effect, the circuits discussed above, although suitable for developing an understanding of the working of the mixed-mode network, need additional mechanisms to impart programmability in the scaling factors. One such technique is discussed next.

4.4 Digitally-Controlled DVCC

Apart from the issues related to chip area and component matching requirements, the use of fixed-valued resistors in the voltage-mode neural circuit for the solution of linear equations presented in Chap. 3 results in a circuit hard-wired for a particular problem. To reduce the hardware complexity, the technique of replacing the voltage-output comparators (implemented using opamps) with current-output comparators (realized using DVCCs) was demonstrated in the previous sections of this chapter, with promising results. However, there was no mechanism in the DVCC-based mixed-mode network to reconfigure a circuit for solving different sets of equations.

Therefore, mechanisms to scale the synaptic currents in accordance with the entries in the coefficient matrix, A (4.6), need to be explored, to incorporate reconfigurability in the circuit. For that purpose, a programmable analog building block viz. Digitally Controlled DVCC (DC-DVCC) is discussed in this section and employed in the next section to yield a mixed-mode linear equation solver with digitally programmable synaptic weights.

The circuit of DC-DVCC obtained after suitable modifications in the circuit of Fig. 4.2 is presented in Fig. 4.6. The technique is to control the current transfer gain parameter k, defined as the ratio of I_{Z+} to I_X, of the DVCC by replacing the Z terminal transistors of the DVCC with transistor arrays associated with switches [10]. The gain parameter k can take values from 1 to $(2^n - 1)$, where n represents the number of transistor arrays. Actually, the transistor arrays implement a current summing network (CSN) at the Z terminal. Although both Z+ and Z− types of current outputs are mentioned in (4.1), the DC-DVCCs used in the network use only Z+ type of outputs. Therefore, only Z+ outputs are shown in Fig. 4.6.

The CSN consists of n transistor pairs, whose NMOS and PMOS aspect ratios are given by:

$$\text{NMOS:} \quad \left(\frac{W}{L}\right)_i = 2^i \left(\frac{W}{L}\right)_{11} \qquad i = 0, 1, 2, \ldots, (n-1) \qquad (4.81)$$

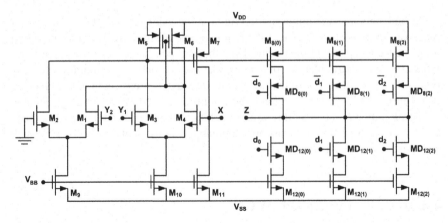

Fig. 4.6 CMOS realization of digitally controlled DVCC with gain k

$$\text{PMOS:} \quad \left(\frac{W}{L}\right)_i = 2^i \left(\frac{W}{L}\right)_7 \quad i = 0, 1, 2, \ldots, (n-1) \quad (4.82)$$

Therefore, the current at the Z terminal, assumed flowing out of the DC-DVCC, can be expressed by

$$I_Z = \sum_{i=0}^{n-1} d_i 2^i (I_7 - I_{11}) \quad (4.83)$$

Therefore, the DC-DVCC provides a current transfer gain equal to:

$$k = \frac{I_Z}{I_X} = \frac{\sum_{i=0}^{n-1} d_i 2^i (I_7 - I_{11})}{(I_7 - I_{11})} = \sum_{i=0}^{n-1} d_i 2^i \quad (4.84)$$

Parameter d_i represents the digital code-bit applied to the i-th branch in the CSN. Depending upon its value, it enables or disables the current to flow in that particular branch. It is instructive to note the numbering of the transistors in the CSN. Transistors labeled $M_{8(i)}$ and $M_{12(i)}$ refer to the PMOS and NMOS transistors in the CSN that have been put there in the place of their counterparts of Fig. 4.2. Transistors $MD_{8(i)}$ and $M_{D12(i)}$ are the actual digital control transistors as the digital control bits d_0, d_1 and d_2 are applied at their respective gate terminals.

As an example, consider a scaling factor of 2 at the $Z+$ terminal. To obtain I_{Z+} at twice the value of I_X, only the second stage (out of the three stages) needs to be operational. Therefore, the digital control word required is [0 1 0], which implies that the first and the third stage are not activated, and the effective Z port current is only due to the second stage comprising of transistors $M_{8(1)}$ and $M_{12(1)}$.

For the DC-DVCC, comparator action can be achieved by putting $R_X \to 0$ i.e. by directly grounding the X-terminal [2]. For such a case, the current in the X-port will saturate and can be written as

$$I_X = I_m \tanh \beta \ (V_{Y1} - V_{Y2}) \tag{4.85}$$

where β is the open-loop gain of the comparator (practically very high), $\pm I_m$ are the saturated output current levels of the comparator and $V_1, V_2, \ldots, V_n$ are the neuron outputs. Equation (4.85) can also be written in an equivalent notation as

$$I_X = g V_m \tanh \beta \ (V_{Y1} - V_{Y2}) \tag{4.86}$$

where g is the transconductance from the input voltage ports (Y_1 and Y_2) to the output current port X and is governed by the resistance at the X port; and $\pm V_m$ are the biasing voltages of the DVCC-based comparator. By virtue of DVCC action, this current will be transferred to the Z+ ports as

$$I_z{}^+ = k I_X = k g V_m \tanh \beta \ (V_{Y1} - V_{Y2}) \tag{4.87}$$

Therefore, the current at the Z terminal can be made a digitally controlled scaled replica of the X port current. This property will be utilized in the design of the multi-output DVCCs needed for the circuit. It may be mentioned that multiple Z+ outputs can be obtained by repeating the output stage comprising of transistors $M_{8(i)}$ and $M_{12(i)}$.

4.5 DC-DVCC Based Linear Equation Solver

The neural-network based circuit to solve the system of equations of (4.5) is presented in Fig. 4.7. As can be seen from Fig. 4.7, individual equations from the set of equations to be solved are passed through non-linear synapses which are realized using multi-output DC-DVCC based voltage comparators with multiple (and programmable) current outputs. The outputs of the comparators are fed to neurons having weighted inputs. These weighted neurons are realized by using opamps where the scaled currents coming from various comparators act as weights. R_{pi} and C_{pi} are the input resistance and capacitance of the opamp corresponding to the ith neuron. These parasitic components are included to model the dynamic nature of the opamp.

To understand the dynamics of the network, we first write the equation of motion of the ith neuron as

$$C_{pi} \frac{du_i}{dt} = k_{1i} I_{X1} + k_{2i} I_{X2} + \ldots + k_{ni} I_{Xn} - \left[\frac{u_i}{R_{pi}} \right] \tag{4.88}$$

where u_i is the internal state of the ith neuron and k_{ji} is the current scaling factor at the jth output of ith DVCC. Using (4.86) in (4.88) results in

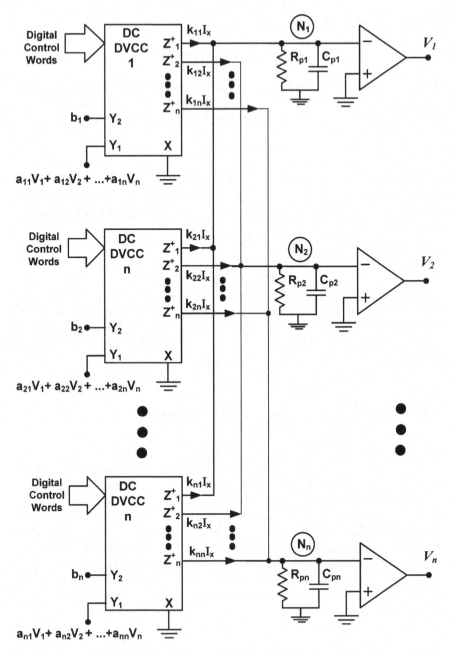

Fig. 4.7 Digitally programmable, non-linear feedback, mixed-mode, neural circuit to solve simultaneous linear equations in n-variables

$$C_{pi} \frac{du_i}{dt} = k_{1i} g V_m \tanh \beta (a_{11} V_1 + a_{12} V_2 + \cdots + a_{1n} V_n - b_1)$$

$$+ k_{2i} g V_m \tanh \beta (a_{21} V_1 + a_{22} V_2 + \cdots + a_{2n} V_n - b_2) + \cdots +$$

$$+ k_{ni} g V_m \tanh \beta (a_{n1} V_1 + a_{n2} V_2 + \cdots + a_{nn} V_n - b_n) - \frac{u_i}{R_{pi}} \quad (4.89)$$

Further, as has been done throughout this text, we associate the network under consideration, with an Energy Function E, which is found by inspection of the equation of motion of the neurons. For the circuit given in Fig. 4.7, a suitable energy function may be

$$E = \frac{V_m}{\beta} \sum_{i=1}^{n} \ln \cosh \beta \left(\sum_{=1}^{n} a_{ij} V_j - b_i \right) - \sum_{i=1}^{n} \frac{1}{R_i} \int_0^{V_i} u_i dV_i \quad (4.90)$$

From (4.90), it follows that

$$\frac{\partial E}{\partial V_i} = V_m a_{1i} \tanh \beta (a_{11} V_1 + a_{12} V_2 + \cdots + a_{1n} V_n - b_1) +$$

$$V_m a_{2i} \tanh \beta (a_{21} V_1 + a_{22} V_2 + \cdots + a_{2n} V_n - b_2) + \cdots +$$

$$V_m a_{ni} \tanh \beta (a_{n1} V_1 + a_{n2} V_2 + \cdots + a_{nn} V_n - b_n) - \frac{u_i}{R_{pi}} \quad (4.91)$$

Also, if 'E' is the Energy Function, it must satisfy the following condition [6].

$$\frac{\partial E}{\partial V_i} = K_R C_{pi} \frac{du_i}{dt} \quad (4.92)$$

where K_R is a constant of proportionality having the dimensions of resistance and is normalized to $1/g$ for simplicity. Comparing (4.89) and (4.91) according to (4.92) yields

$$\begin{bmatrix} k_{11} & k_{12} & \cdots & k_{1n} \\ k_{21} & k_{22} & \cdots & k_{2n} \\ \vdots & \vdots & \cdots & \vdots \\ k_{n1} & k_{n2} & \cdots & k_{nn} \end{bmatrix} = \begin{bmatrix} a_{11} & a_{12} & \cdots & a_{1n} \\ a_{21} & a_{22} & \cdots & a_{2n} \\ \vdots & \vdots & \cdots & \vdots \\ a_{n1} & a_{n2} & \cdots & a_{nn} \end{bmatrix} \quad (4.93)$$

It needs to be mentioned here that the complexity of the NOSYNN based digitally programmable mixed-mode linear equation solver presented here is further reduced from its voltage-mode counterpart (which itself compares favorably with existing networks), with the DC-DVCC based neural circuit for solving linear equations employing n operational amplifiers, n multi-output DC-DVCCs and $(n^2 + n)$ resistances. The reduction in complexity has been possible by virtue of the use of DC-DVCC based voltage comparators with current outputs thereby causing the synaptic signals to be conveyed as currents and eliminating the need of synaptic resistances.

However, as in the case of the previously discussed circuit, some additional resistances are required to generate the voltages to be applied at the inputs of the voltage comparators. The method for obtaining the values of these resistances remains the same as explained before.

4.6 Hardware Simulation Results

The operation of the network of Fig. 4.7 was verified through computer simulations using PSPICE program by solving various sets of simultaneous linear equations in 2, 3, 4, 5 and 10 variables.

PSPICE Simulation Result: 2 linear equations in 2 variables
A simple circuit comprising of 2 neurons (realized using operational amplifiers) and 2 non-linear synapses (comprising of DC-DVCC based voltage comparators with multiple $Z+$ outputs) was first tested using computer simulations using the PSPICE program. Consider the following system of linear equations in 2 variables:

$$\begin{bmatrix} 1 & 2 \\ 2 & 1 \end{bmatrix} \begin{bmatrix} V_1 \\ V_2 \end{bmatrix} = \begin{bmatrix} 3.5 \\ 4 \end{bmatrix} \tag{4.94}$$

The mathematical solution of (4.94) is $V_1 = +1.5$ and $V_2 = +1$. The values of the current transfer gains required to be set for the two outputs of the two DC-DVCCs can be found as

$$k_{11} = a_{11} = 1$$
$$k_{12} = a_{12} = 2$$
$$k_{21} = a_{21} = 2$$
$$k_{22} = a_{22} = 1$$

which correspond to the following digital control words

$$DC - DVCC - 1, \ Z_1+ \ \rightarrow \ [001]$$
$$DC - DVCC - 1, \ Z_2+ \ \rightarrow \ [010]$$
$$DC - DVCC - 2, \ Z_1+ \ \rightarrow \ [010]$$
$$DC - DVCC - 2, \ Z_1+ \ \rightarrow \ [001]$$

The scaling factors given above may be implemented in the DC-DVCC by proper selecting the digital control words for the $Z+$ stages of the DVCC based comparators. For instance, for the first DVCC, the second $Z+$ stage control word should be [0 1 0], and the control word for the second $Z+$ stage of the second DC-DVCC should be kept as [0 0 1]. During the course of PSPICE simulations, a 'high' on a digital control bit was provided by connecting the gates of the NMOS and PMOS digital control transistors to V_{DD} and V_{SS} respectively. A 'low' i.e. '0' can be implemented by reversing these connections.

As has already been explained in the previous chapter, some additional circuitry is needed to generate the voltage required at the input of the comparators. This can be done by a resistive potential divider, as is explained next. We shall consider a nomenclature similar to that used in the case of the voltage-mode linear equation solver. To obtain the values of the resistances required at the input of the first comparator, we proceed by properly scaling the first equation from the set of linear Eq. (4.94). Selecting a scaling factor of 3, we have

$$\frac{1}{3}V_1 + \frac{2}{3}V_2 = \frac{3.5}{3} \tag{4.95}$$

from where the values of the resistances to be used in the potential divider can be obtained as

$$R_{e11} = 3\,K\Omega$$
$$R_{e12} = 1.5\,K\Omega$$
$$R_{e13} = \infty$$

and the voltage to be applied at the Y_2 input terminal of the first DVCC based comparator would be

$$b_1 = \frac{3.5}{3} = 1.166\,V$$

Similarly, scaling the second equation by a factor of 4, we have

$$\frac{2}{4}V_1 + \frac{1}{4}V_2 = \frac{4}{4} \tag{4.96}$$

from where the values of the resistances to be used in the potential divider can be obtained as

$$R_{e21} = 2\,K\Omega$$
$$R_{e22} = 4\,K\Omega$$
$$R_{e23} = 4\,K\Omega$$

and the voltage to be applied at the Y_2 input terminal of the second DVCC based comparator would be

$$b_2 = \frac{4}{4} = 1\,V$$

The mixed-mode circuit was tested in PSPICE and the steady state neuronal output voltages were obtained as $V_1 = 1.51\,V$ and $V_2 = 1.02\,V$, which are in close agreement with the mathematical results for the chosen set of linear equations.

PSPICE Simulation Result: 3 linear equations in 3 variables

The application of the circuit to a chosen 3-variable problem (4.97) is presented below.

$$\begin{bmatrix} 2 & 1 & 1 \\ 3 & 2 & 1 \\ 1 & 1 & 2 \end{bmatrix} \begin{bmatrix} V_1 \\ V_2 \\ V_3 \end{bmatrix} = \begin{bmatrix} 5 \\ 10 \\ 6 \end{bmatrix} \tag{4.97}$$

The circuit to solve (4.97) as obtained from Fig. 4.7 is presented in Fig. 4.8. The values of the current scaling coefficients k_{ji} are given as

$$\begin{bmatrix} k_{11} & k_{12} & k_{13} \\ k_{21} & k_{22} & k_{23} \\ k_{31} & k_{32} & k_{33} \end{bmatrix} = \begin{bmatrix} 2 & 1 & 1 \\ 3 & 2 & 1 \\ 1 & 1 & 2 \end{bmatrix} \tag{4.98}$$

which correspond to the following digital control words

$$DC - DVCC - 1, \; Z_1 + \; \rightarrow [010]$$
$$DC - DVCC - 1, \; Z_2 + \; \rightarrow [001]$$
$$DC - DVCC - 1, \; Z_3 + \; \rightarrow [001]$$
$$DC - DVCC - 2, \; Z_1 + \; \rightarrow [011]$$
$$DC - DVCC - 2, \; Z_2 + \; \rightarrow [010]$$
$$DC - DVCC - 2, \; Z_3 + \; \rightarrow [001]$$
$$DC - DVCC - 3, \; Z_1 + \; \rightarrow [001]$$
$$DC - DVCC - 3, \; Z_2 + \; \rightarrow [001]$$
$$DC - DVCC - 3, \; Z_3 + \; \rightarrow [010]$$

As can be seen from Fig. 4.8, some additional circuitry is needed to generate the inputs to the non-linear synapses. Routine analysis yields the following values of the resistors at the input of the non-linear synapses.

$$R_{e11} = 2.5K, \, R_{e12} = R_{e13} = R_{e14} = 5K,$$

$$R_{e21} = 3.33K, \, R_{e22} = 5K, \, R_{e23} = 10K, \, R_{e24} = 2.5K,$$

$$R_{e31} = R_{e32} = 6K, \, R_{e33} = R_{e34} = 3K,$$

$$R_{c11} = R_{c21} = R_{c31} = 1K,$$

$$R_{c12} = 4K, \, R_{c22} = 9K, \, R_{c32} = 5K.$$

Algebraic analysis of (4.97) gives the solution as $V_1 = -0.5$ V, $V_2 = 5.5$ V and $V_3 = 0.5$ V. The results of PSPICE simulation of the circuit of Fig. 4.8, shown in Fig. 4.9, are found to match perfectly with the algebraic solution. The initial node voltages were kept as $V(1) = 10 \, mV$, $V(2) = -10 \, mV$ and $V(3) = 20 \, mV$. The circuit for DVCC was taken from [2] and standard 0.5 micron CMOS parameters were used for simulation purposes. The DVCCs were biased with $\pm 2.5V$ supplies. For the opamp, use was made of the LMC7101A CMOS opamp from National Semiconductor. The sub-circuit file for this opamp is available in Orcad Model Library.

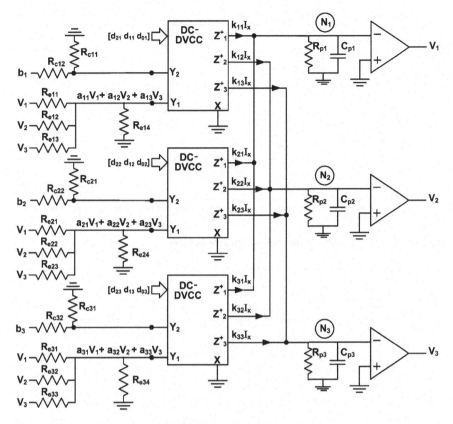

Fig. 4.8 The DC-DVCC based circuit applied to a 3-variable problem

The biasing voltages for the opamps were taken to be ± 15V. The transfer characteristics of the DVCC-based comparator, shown in Fig. 4.10, were also plotted using SPICE simulation and value of 'β' was found to be 2.4×10^3.

The circuit was further tested in PSPICE for solving systems of linear equations in 4, 5, and 10 variables. Results of simulation runs for these problems are presented in Table 4.3. Each of the simulations was run using various initial conditions in the millivolt range. As can be seen from Table 4.3, the network always converges to the solution of the given system of linear equations, and the obtained results are quite near the algebraic solutions with the maximum error being approximately 6 % and the average error being 0.49 %.

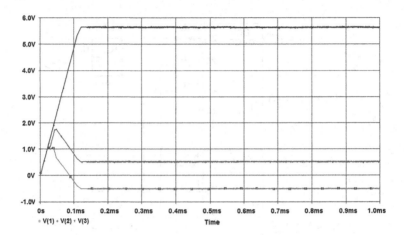

Fig. 4.9 Simulation results for the chosen 3-variable problem

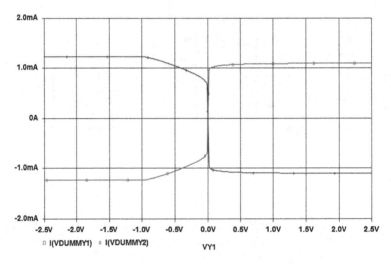

Fig. 4.10 Transfer characteristics of the DC-DVCC used to realize the comparator

4.7 Performance Evaluation

This section deals with the effect of component mismatches and device non-idealities on the accuracy of the solution obtained from the circuit. If the circuit is implemented using discrete resistances, variations in the values of the resistances need to be considered keeping in mind the tolerances associated with discrete resistors. However, if the network is targeted for monolithic integration, *random* variations in the resistance values need not be considered. In that case, all the resistances are expected to deviate from their assigned values by the same factor [11]. Therefore, a worthy performance appraisal of the network could be obtained by testing the circuit with all resistances

Table 4.3 PSPICE simulation results for the DC-DVCC based linear equation solver circuit applied to solve different systems of linear equations

[A]	[B]	Algebraic solution [V]	Simulated results (using PSPICE) [V]	Percentage error in the solution (%)
$\begin{bmatrix} 1 & 2 \\ 2 & 1 \end{bmatrix}$	$\begin{bmatrix} 3.5 \\ 4 \end{bmatrix}$	$\begin{bmatrix} 1.5 \\ 1 \end{bmatrix}$	$\begin{bmatrix} 1.51 \\ 1.02 \end{bmatrix}$	$\begin{bmatrix} 0.66 \\ 2.00 \end{bmatrix}$
$\begin{bmatrix} 2 & 1 & 1 \\ 3 & 2 & 1 \\ 1 & 1 & 2 \end{bmatrix}$	$\begin{bmatrix} 5 \\ 10 \\ 6 \end{bmatrix}$	$\begin{bmatrix} -0.5 \\ 5.5 \\ 0.5 \end{bmatrix}$	$\begin{bmatrix} -0.51 \\ 5.54 \\ 0.48 \end{bmatrix}$	$\begin{bmatrix} 4.00 \\ 0.72 \\ -4.00 \end{bmatrix}$
$\begin{bmatrix} 2 & 3 & 6 & 1 \\ 3 & 2 & 5 & 4 \\ 2 & 5 & 5 & 2 \\ 2 & 4 & 2 & 5 \end{bmatrix}$	$\begin{bmatrix} 48.75 \\ 67.5 \\ 54 \\ 59.25 \end{bmatrix}$	$\begin{bmatrix} 3.5 \\ 1.5 \\ 5 \\ 7.25 \end{bmatrix}$	$\begin{bmatrix} 3.51 \\ 1.53 \\ 5.09 \\ 7.19 \end{bmatrix}$	$\begin{bmatrix} 0.28 \\ 2.00 \\ 1.80 \\ -0.83 \end{bmatrix}$
$\begin{bmatrix} 2 & 3 & 9 & 2 & 5 \\ 2 & 6 & 9 & 9 & 5 \\ 2 & 6 & 2 & 4 & 5 \\ 2 & 4 & 7 & 8 & 3 \\ 5 & 3 & 6 & 3 & 5 \end{bmatrix}$	$\begin{bmatrix} -54.9 \\ -91.0 \\ -54.9 \\ -74.8 \\ -70.0 \end{bmatrix}$	$\begin{bmatrix} -6.7 \\ -4.3 \\ -2.8 \\ -3.3 \\ 0.64 \end{bmatrix}$	$\begin{bmatrix} -6.55 \\ -4.33 \\ -2.63 \\ -3.19 \\ 0.63 \end{bmatrix}$	$\begin{bmatrix} -0.75 \\ 0.69 \\ -6.07 \\ -3.33 \\ -1.56 \end{bmatrix}$
$\begin{bmatrix} 1 & 2 & 1 & 3 & 4 & 2 & 1 & 1 & 1 & 2 \\ 2 & 1 & 1 & 3 & 1 & 2 & 2 & 1 & 2 & 3 \\ 1 & 1 & 4 & 3 & 3 & 1 & 1 & 4 & 4 & 1 \\ 4 & 2 & 1 & 5 & 3 & 3 & 1 & 1 & 2 & 2 \\ 1 & 1 & 5 & 1 & 2 & 1 & 2 & 2 & 5 & 2 \\ 3 & 3 & 1 & 2 & 1 & 1 & 5 & 2 & 1 & 1 \\ 5 & 5 & 1 & 4 & 1 & 1 & 3 & 4 & 2 & 2 \\ 1 & 1 & 1 & 2 & 2 & 2 & 3 & 3 & 4 & 1 \\ 4 & 2 & 2 & 1 & 3 & 2 & 5 & 4 & 3 & 2 \\ 1 & 2 & 3 & 1 & 2 & 3 & 1 & 2 & 3 & 4 \end{bmatrix}$	$\begin{bmatrix} 10 \\ -11 \\ 10 \\ 2 \\ -7 \\ -9 \\ -8 \\ -3 \\ -3 \\ 7 \end{bmatrix}$	$\begin{bmatrix} -2 \\ 1 \\ 3 \\ -1 \\ 2 \\ 7 \\ -3 \\ 4 \\ -5 \\ -4 \end{bmatrix}$	$\begin{bmatrix} -1.91 \\ 1.04 \\ 3.05 \\ -0.95 \\ 1.94 \\ 7.04 \\ -3.06 \\ 3.99 \\ -4.91 \\ -3.95 \end{bmatrix}$	$\begin{bmatrix} -4.50 \\ 4.00 \\ 1.66 \\ -5.00 \\ -3.00 \\ 0.57 \\ 2.00 \\ -0.25 \\ -1.80 \\ -1.25 \end{bmatrix}$

having the same percentage deviation from their assigned values. Such an assessment of the quality of the obtained solution is presented in Table 4.4 from which it can be seen that the percentage error in the solution varies with the percentage errors specified in the values of the resistances and even for $\pm 10\%$ variation in the values of the resistors, the solution point changes by approximately 3 %.

Further, the gains of the DVCC-based comparators were varied to investigate their effect on the solution quality. Toward that end, a resistance was connected between the X-terminal of each DVCC and ground. Ideally, these resistances were assumed to have zero value. However, by assigning different values to these resistances, the gains of the DVCCs were varied. The solutions, as obtained for the 3-variable problem of Fig. 4.8, are presented in Table 4.5 below. It can be seen that small variations in the gains of the comparators do not affect the quality of the solution.

Next, the effect of offset voltages in the DVCC-based comparators was explored. Offset voltages were applied at the Y_2 inputs of the DVCCs of Fig. 4.8 and the results of PSPICE simulations were compared with the algebraic solution as given

Table 4.4 Effect of variation in resistances on the obtained results

Percentage variation in resistances (%)	Simulated results (using PSPICE) [V]	Percentage error in the solution (%)
+2	$\begin{bmatrix} -0.508 \\ 5.602 \\ 0.499 \end{bmatrix}$	$\begin{bmatrix} +1.6 \\ +1.85 \\ -0.2 \end{bmatrix}$
+5	$\begin{bmatrix} -0.509 \\ 5.619 \\ 0.507 \end{bmatrix}$	$\begin{bmatrix} +1.8 \\ +2.16 \\ +0.2 \end{bmatrix}$
+10	$\begin{bmatrix} -0.514 \\ 5.681 \\ 0.507 \end{bmatrix}$	$\begin{bmatrix} +2.8 \\ +3.29 \\ +1.4 \end{bmatrix}$
−2	$\begin{bmatrix} -0.507 \\ 5.601 \\ 0.505 \end{bmatrix}$	$\begin{bmatrix} +1.4 \\ +1.83 \\ +1 \end{bmatrix}$
−5	$\begin{bmatrix} -0.518 \\ 5.679 \\ 0.506 \end{bmatrix}$	$\begin{bmatrix} +3.6 \\ +3.25 \\ +1.2 \end{bmatrix}$
−10	$\begin{bmatrix} -0.516 \\ 5.678 \\ 0.504 \end{bmatrix}$	$\begin{bmatrix} +3.2 \\ +3.23 \\ +0.8 \end{bmatrix}$

Table 4.5 Effect of gains of the DVCC-based comparators on the solution quality

X-terminal resistances of the DVCCs (Ω)	Simulated results (using PSPICE) [V]	Percentage error in the solution (%)
5	$\begin{bmatrix} -0.512 \\ 5.610 \\ 0.500 \end{bmatrix}$	$\begin{bmatrix} 2.4 \\ 2.0 \\ 0.0 \end{bmatrix}$
10	$\begin{bmatrix} -0.514 \\ 5.617 \\ 0.510 \end{bmatrix}$	$\begin{bmatrix} 2.8 \\ 2.1 \\ 2.00 \end{bmatrix}$
20	$\begin{bmatrix} -0.520 \\ 5.620 \\ 0.511 \end{bmatrix}$	$\begin{bmatrix} -4.00 \\ 2.18 \\ 2.2 \end{bmatrix}$

in Table 4.6. As can be seen, the offset voltages of the comparators do not affect the obtained solutions to any appreciable extent. However, the error does tend to increase with increasing offset voltages.

Finally, offset voltages for the opamps were also considered. Offset voltages were applied at the non-inverting inputs of opamps of Fig. 4.8 and the results of PSPICE simulations were compared with algebraic solution as given in Table 4.7. As can be seen, the offset voltages of the operational amplifiers have little effect on the obtained solutions, with the maximum error of about 3.5 % arising for a +15 mV offset.

Table 4.6 Effect of offset voltages of the DVCC-based comparators on the solution quality	Offset voltage applied at Y_2 input of the DVCCs (mV)	Simulated results (using PSPICE) [V]	Percentage error in the solution (%)
	-5	$\begin{bmatrix} -0.498 \\ 5.615 \\ 0.497 \end{bmatrix}$	$\begin{bmatrix} -0.4 \\ 2.09 \\ -0.6 \end{bmatrix}$
	-10	$\begin{bmatrix} -0.505 \\ 5.620 \\ 0.506 \end{bmatrix}$	$\begin{bmatrix} 1.0 \\ 2.18 \\ 1.2 \end{bmatrix}$
	-15	$\begin{bmatrix} -0.512 \\ 5.631 \\ 0.511 \end{bmatrix}$	$\begin{bmatrix} 2.4 \\ 2.38 \\ 2.2 \end{bmatrix}$
	$+5$	$\begin{bmatrix} -0.496 \\ 5.611 \\ 0.494 \end{bmatrix}$	$\begin{bmatrix} -0.8 \\ 2.01 \\ -1.2 \end{bmatrix}$
	$+10$	$\begin{bmatrix} -0.503 \\ 5.627 \\ 0.508 \end{bmatrix}$	$\begin{bmatrix} 0.6 \\ 2.30 \\ 1.6 \end{bmatrix}$
	$+15$	$\begin{bmatrix} -0.519 \\ 5.629 \\ 0.510 \end{bmatrix}$	$\begin{bmatrix} 3.8 \\ 2.34 \\ 2 \end{bmatrix}$

4.8   VLSI Implementation Issues

Proper selection of the control words can be used to achieve any current scaling factor between 1 and 7, and therefore the realizations of mixed-mode digitally programmable networks discussed in this chapter, are restricted to solving systems of equations with coefficients in that range. However, the range may be widened further by adding more stages in the CSN. For instance, a 4 bit CSN would allow the current scaling factors to be varied between 0 to 15. The limit on the number of stages which can be incorporated in the CSN is set by the transistor sizing requirements. As each digitally controlled stage which is added to the CSN has transistors area which is double the previous stage, the overall area requirement tends to shoot up for a large number of stages. For instance, if the fifth bit is to be added in the digital control word, it would mean adding transistors whose areas are 32 times the areas of their counterparts in the X stage.

It needs to be mentioned that the DVCC implementation presented in Fig. 4.2 is not the only CMOS realization available in the technical literature, and a multitude of different realizations have been proposed over the past [4, 12, 13]. The CMOS implementation of Fig. 4.2 was chosen due to its relatively better current conveying characteristics and simpler circuit structure. The reader is encouraged to verify the

Table 4.7 Effect of offset voltages of opamps on solution quality

Offset voltage applied at the non-inverting input of opamps (mV)	Simulated results (using PSPICE) [V]	Percentage error in the solution (%)
−5	$\begin{bmatrix} -0.494 \\ 5.617 \\ 0.507 \end{bmatrix}$	$\begin{bmatrix} -1.2 \\ 2.12 \\ 1.4 \end{bmatrix}$
−10	$\begin{bmatrix} -0.504 \\ 5.581 \\ 0.510 \end{bmatrix}$	$\begin{bmatrix} 0.8 \\ 1.47 \\ 2 \end{bmatrix}$
−15	$\begin{bmatrix} -0.514 \\ 5.543 \\ 0.491 \end{bmatrix}$	$\begin{bmatrix} 2.8 \\ 0.78 \\ -1.8 \end{bmatrix}$
+5	$\begin{bmatrix} -0.499 \\ 5.642 \\ 0.497 \end{bmatrix}$	$\begin{bmatrix} -0.2 \\ 2.58 \\ -0.6 \end{bmatrix}$
+10	$\begin{bmatrix} -0.506 \\ 5.668 \\ 0.505 \end{bmatrix}$	$\begin{bmatrix} 1.2 \\ 3.05 \\ 1.0 \end{bmatrix}$
+15	$\begin{bmatrix} -0.512 \\ 5.693 \\ 0.507 \end{bmatrix}$	$\begin{bmatrix} 2.4 \\ 3.51 \\ 1.4 \end{bmatrix}$

performance of the circuits presented in this chapter by choosing an alternative CMOS realization of the DVCC.

Another pertinent issue is the technology node at which the CMOS realizations of the DVCC and the DC-DVCC are discussed. The $0.5\,\mu$m CMOS process parameters are a thing of the past, with modern day integration process achieving transistor sizes of the order of tens of nanometers. Therefore, it needs to be pointed out that a more realistic picture could be obtained by re-simulating the various circuits after employing a DVCC comprising of MOSFETs having aspect ratios which correspond to present day technologies.

The resistances required to generate the voltages at the inputs of the various DC-DVCC based comparators can be eliminated by replacing the operational amplifiers (emulating the functionality of neurons) by multiple output DC-DVCCs. Scaled current available from such DC-DVCCs may then be fed back to the inputs of the DC-DVCC based comparators. However, the grounded resistance shown at the inputs of the comparators would still need to be connected for the purpose of converting the currents arriving at that node, to a corresponding voltage. If such an approach is followed, the need for scaling of the equations is virtually eliminated.

Lastly, it should be mentioned that a multi-output Operational Transconductance Amplifier (MO-OTA) could also be used in place of the multi-output DVCCs for implementing the voltage-comparators. In that case, the transconductance gain, g_m, is governed by the bias current of the device. For obtaining characteristics similar to the tanh(.) function, the bias current should be kept as high as possible in the particular realization of the OTA used.

4.9 Conclusion

In this chapter, a mixed-mode approach to solve n simultaneous linear equations in n variables, which uses n neurons and n synapses, has been described. Each neuron requires one opamp and each synapse is implemented using one comparator. The comparators were realized using a digitally controlled DVCC. From VLSI implementation point of view, use of CMOS DC-DVCCs and opamps facilitates viability for monolithic integration. The working of the network was verified using PSPICE for various sample problem sets of 2 to 10 simultaneous linear equations. While the simulation results confirm the validity of the approach, issues such as the network response to a system of equations which do *not* have a unique solution are yet to be explored. It may be mentioned that the technique can be extended to the standard linear and quadratic programming problems, as shall be discussed in a later chapter. Similarly, the circuit complexity of the modified NOSYNN based voltage-mode graph colouring network of Chap. 2 can be reduced by employing DVCC-based voltage comparators with current outputs. The resulting mixed-mode neural circuit for graph colouring is discussed in Appendix A at the end of the book.

References

1. Elwan, H.O., Soliman, A.M.: Novel CMOS differential voltage current conveyor and its applications. IEE Proc. Circ. Dev. Syst. **144**(3), 195–200 (1997)
2. Maheshwari, S.: A canonical voltage-controlled VM-APS with a grounded capacitor. Circ. Syst. Sig. Proces. **27**(1), 123–132 (2008)
3. Soliman, A.M.: Generation and classification of Kerwin-Huelsman-Newcomb circuits using the DVCC. Int. J. Circ. Theory Appl. **37**, 835–855 (2008)
4. Hassan, T.M., Mahmoud, S.A.: New CMOS DVCC realization and applications to instrumentation amplifier and active-RC filters. Int. J. Electron. Commun. **64**, 47–55 (2010)
5. Khateba, F., Khatiba, N., Koton, J.: Novel low-voltage ultra-low-power DVCC based on floating-gate folded cascode OTA. Microelectron. J. **42**(8), 1010–1017 (2011)
6. Rahman, S.A., Jayadeva, Dutta Roy, S.C.: Neural network approach to graph colouring. Electron. Lett. **35**(14), 1173–1175 (1999)
7. Rahman, S.A.: A nonlinear synapse neural network and its applications. PhD thesis, Department of Electrical Engineering, Indian Institute of Technology, Delhi, India (2007)
8. Tank, D., Hopfield, J.: Simple 'neural' optimization networks: an A/D converter, signal decision circuit, and a linear programming circuit. IEEE Trans. Circ. Syst. **33**(5), 533–541 (1986)
9. LMC7101A. National semiconductor inc. http://www.national.com/assets/en/tools/spice/LMC7101A.MOD. Last Accessed on 25 Oct 2012
10. Hassan, T.M., Mahmoud, S.A.: Fully programmable universal filter with independent gain-ω_o − Q control based on new digitally programmable CMOS CCII. J. Circ. Syst. Comput. **18**(5), 875–897 (2009)

11. Plummer, J.D., Deal, M.D., Griffin, P.B.: Silicon VLSI Technology: Fundamentals, Practice and Modeling. Prentice Hall, Upper Saddle River (2000)
12. Hu, J., Xu, T., Zhang, W., Xia, Y.: A CMOS rail-to-rail differential voltage current conveyor and its applications. In: Communications, Circuits and Systems, 2005. Proceedings. 2005 International Conference on, vol. 2. IEEE (2005)
13. Mahmoud, S.A.: Low voltage wide range CMOS differential voltage current conveyor and its applications. Contemp. Eng. Sci. **1**(3), 105–126 (2008)

Chapter 5
Non-Linear Feedback Neural Circuits for Linear and Quadratic Programming

5.1 Introduction

Mathematical programming, in general, is concerned with the determination of a minimum or a maximum of a function of several variables, which are required to satisfy a number of constraints. Such solutions are sought in diverse fields including engineering, operations research, management science, computer science, numerical analysis, and economics [1, 2].

A general mathematical programming problem can be stated as [2]:

$$Minimize \ \ f(x)$$

subject to

$$g_i \left(\mathbf{x} \right) \geq 0 \ \ (i = 1, 2, \ldots, m) \tag{5.1}$$

$$h_j \left(\mathbf{x} \right) = 0 \ \ (j = 1, 2, \ldots, p) \tag{5.2}$$

$$\mathbf{x} \in S \tag{5.3}$$

where $\mathbf{x} = (x_1, x_2, \ldots, x_n)^T$ is the vector of unknown decision variables, and $f, g_i (i = 1, 2, \ldots, m), h_j (j = 1, 2, \ldots, p)$ are the real-valued functions of the n real variables $x_1, x_2, \ldots, x_n$.

In this formulation, the function f is called the *objective function*, and inequalities (5.1), Eq. (5.2) and the set restrictions (5.3) are referred to as the *constraints*. It may be mentioned that although the mathematical programming problem (MPP) has been stated as a minimization problem in the description above, the same may readily be converted into a maximization problem without any loss of generality, by using the identity given in (5.4)

$$\max f (x) = -\min [-f (x)] \tag{5.4}$$

M. S. Ansari, *Non-Linear Feedback Neural Networks*, 145
Studies in Computational Intelligence 508, DOI: 10.1007/978-81-322-1563-9_5,
© Springer India 2014

When all the functions appearing in the MPP are linear in the decision variables **x**, the problem is referred to as a *linear programming problem* (LPP). If however, $f(x)$ is a second-order function of $x_1, x_2, \ldots, x_n$, then the problem is called a *quadratic programming problem* (QPP).

Traditional methods for solving linear and quadratic programming problems typically involve an iterative process, but long computational time limits their usage, because traditional algorithms for digital computers may not be efficient since the computing time required for a solution is greatly dependent on the dimension and structure of the problems [3]. An alternative approach to solution of this problem is to exploit neural networks which have been applied to several classes of constrained optimization problems and have shown promise for solving such problems more effectively. For example, the Hopfield neural network has proved to be a useful tool for solving some of the optimization problems. Tank and Hopfield first NOSYNN-based a neural network for solving mathematical programming problems, where a linear programming problem (LPP) was mapped into a closed-loop network [4]. Since then, various neural networks for solving LPP and QPP have been proposed.

In this chapter, hardware solutions to the problem of solving linear and quadratic programming problems are presented. The NOSYNN-based architecture is essentially an extension of the voltage-mode neural network of Chap. 3. Being based on the NOSYNN, the architecture employs non-linear feedback which leads to a new energy function that involves transcendental terms. This transcendental energy function is fundamentally different from the standard quadratic form associated with Hopfield network and its variants. To solve a LPP in n variables with m linear constraints, the circuit requires n opamps, m unipolar comparators and $(m + n)$ resistances thereby causing the hardware complexity of the NOSYNN-based network to compare favorably with the existing hardware implementations. Similarly, to solve a QPP, in n variables with m linear constraints, the circuit requires n opamps, m unipolar comparators and $(n^2 + mn)$ resistors.

This chapter is organized as follows. Section 5.2 contains the details of the NOSYNN-based neural network for solving LPP along with the design equations. Circuit implementation of the NOSYNN-based network for various sample problems are discussed and PSPICE simulation results of the NOSYNN-based circuit are also presented. Section 5.3 presents the details of the NOSYNN-based neural network for solving QPP along with the results of PSPICE simulations for the NOSYNN-based circuit applied to solve a chosen problem. Section 5.4 deals with the explanation of the energy functions associated with the NOSYNN-based LPP and QPP solvers and the proof of their validity. Issues that are expected to arise in actual monolithic implementations are discussed in Sect. 5.5. Concluding remarks are presented in Sect. 5.6.

5.2 Non-Linear Feedback Neural Network for Linear Programming

Let the first-order function to be minimized be

$$
F = \begin{bmatrix} c_1 & c_2 & \ldots & c_n \end{bmatrix} \begin{bmatrix} V_1 \\ V_2 \\ \vdots \\ V_n \end{bmatrix} \tag{5.5}
$$

subject to the following linear constraints

$$
\begin{bmatrix} a_{11} & a_{12} & \ldots & a_{1n} \\ a_{21} & a_{22} & \ldots & a_{2n} \\ \vdots & \vdots & \ddots & \vdots \\ a_{m1} & a_{m2} & \ldots & a_{mn} \end{bmatrix} \begin{bmatrix} V_1 \\ V_2 \\ \vdots \\ V_n \end{bmatrix} \leq \begin{bmatrix} b_1 \\ b_2 \\ \vdots \\ b_m \end{bmatrix} \tag{5.6}
$$

where $V_1, V_2, \ldots, V_n$ are the variables, and a_{ij}, c_j and b_i ($i = 1, 2, \ldots, m$; $j = 1, 2, \ldots, n$) are constants. The NOSYNN-based neural-network based circuit to minimize the function given in (5.5) in accordance with the constraints of (5.6) is presented in Fig. 5.1. As can be seen from Fig. 5.1, individual inequalities from the set of constraints are passed through non-linear synapses which are realized using unipolar comparators comprising of operational amplifiers and diodes. The outputs of the comparators are fed to neurons having weighted inputs. The neurons are realized by using opamps and the weights are implemented using resistances. The currents arriving to the neuron from various synapses get added up at the input of the neuron. R_{pi} and C_{pi} are the input resistance and capacitance of the opamp that is used to emulate the functionality of a neuron. These parasitic components are included to model the dynamic nature of the operational amplifier.

Graphical representation of the transfer characteristics for a bipolar comparator is shown in Fig. 5.2a from where it can be seen that the comparator output saturates at $\pm V_m$ when the two inputs differ by more than few milli-volts in magnitude. Unipolar transfer characteristics can be obtained using an opamp (the output of which can be modelled by (5.7)) by employing a diode as depicted in Fig. 5.3, the diode essentially 'trimming' one half of the transfer characteristic curve, which are shown in Fig. 5.2b and can be mathematically modelled by (5.8). As is explained in Sect. 5.6, such unipolar comparator characteristics are utilized to obtain an energy function which acts to bring the neuronal states to the *feasible* region.

$$
x = V_m \tanh \beta \left(V_i - V_j \right) \tag{5.7}
$$

$$
x = \frac{1}{2} V_m \left[\tanh \beta \left(V_i - V_j \right) + 1 \right] \tag{5.8}
$$

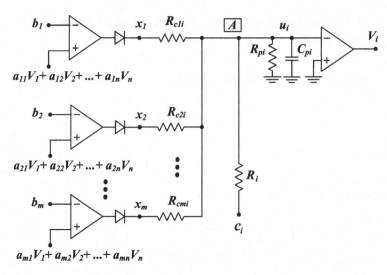

Fig. 5.1 ith neuron of the NOSYNN-based feedback neural network circuit to solve a linear programming problem in n variables with m linear constraints

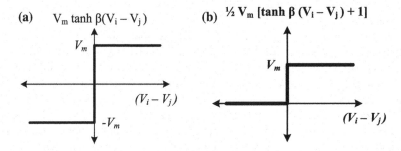

Fig. 5.2 Transfer characteristics of **a** bipolar; and **b** unipolar; comparators

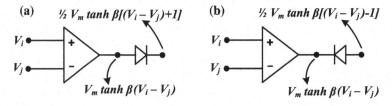

Fig. 5.3 Obtaining unipolar comparator characteristics using an opamp and a diode

Using (5.8), the output of the ith unipolar comparator in Fig. 5.1 can be given by (5.9) where β is the open-loop gain of the comparator (practically very high), $\pm V_m$ are the output voltage levels of the comparator and V_1, V_2, ..., V_n are the neuron outputs.

$$x_i = \frac{1}{2} V_m \ [\tanh \beta \ (a_{i1}V_1 + a_{i2}V_2 + \cdots + a_{in}V_n) + 1] \tag{5.9}$$

Applying node equations for node 'A' in Fig. 5.1, the equation of motion of the ith neuron can be given as

$$C_i \frac{du_i}{dt} = \sum_{j=1}^{m} \left[\frac{x_j}{R_{cji}} \right] + \frac{c_i}{R_i} - \frac{u_i}{R_{ieqv}} \tag{5.10}$$

where R_{ieqv} is the parallel equivalent of all resistances connected at node 'A' in Fig. 5.1 and is given by (5.11)

$$\frac{1}{R_{ieqv}} = \sum_{j=1}^{m} \left[\frac{1}{R_{cji}} \right] + \frac{1}{R_i} + \frac{1}{R_{pi}} \tag{5.11}$$

where u_i is the internal state of the ith neuron, R_{c1i}, R_{c2i}, ..., R_{cmi} are the weight resistance connecting the outputs of the unipolar comparators to the input of the ith neuron. As is shown later in this section, the values of these resistances are governed by the entries in the coefficient matrix of (5.6). Resistance R_i causes terms corresponding to the linear function to be minimized, to appear in the energy function. Further, as has been explained in Sect. 5.6, the non-linear feedback neural circuit of Fig. 5.1 is associated with the following Lyapunov function (also referred to as the 'energy function')

$$E = \sum_{i=1}^{n} c_i V_i + \frac{V_m}{2} \sum_{i=1}^{m} \sum_{j=1}^{n} a_{ij} V_j$$

$$+ \frac{V_m}{2\beta} \sum_{i=1}^{m} \ln \ \cosh \beta \left[\sum_{j=1}^{n} (a_{ij} V_j - b_i) \right] \tag{5.12}$$

This expression of the Energy Function can be written in a slightly different (but more illuminating) form as

$$E = \sum_{i=1}^{n} c_i V_i + (P_1 + P_2 + \cdots + P_m) \tag{5.13}$$

where the first term is the same as the first-order function to be minimized, as given in (5.5), and P_1, P_2, ..., P_m are the penalty terms. The ith penalty term can be given as

$$P_i = \frac{V_m}{2} \sum_{j=1}^{n} a_{ij} V_j + \frac{V_m}{2\beta} \ln \ \cosh \beta \left[\sum_{j=1}^{n} (a_{ij} V_j - b_i) \right] \tag{5.14}$$

Obtaining a partial differentiation of the combined penalty term, $P \ (= P_1 + P_2 + \cdots + P_m)$ with respect to V_i we have

$$\frac{\partial P}{\partial V_i} = \frac{V_m}{2} \sum_{j=1}^{m} a_{ji} + \frac{V_m}{\beta} \sum_{j=1}^{m} a_{ji} \tanh \beta \left[\sum_{j=1}^{n} (a_{ij} V_j - b_i) \right] \qquad (5.15)$$

which may be simplified to

$$\frac{\partial P}{\partial V_i} = \sum_{j=1}^{m} a_{ji} x_j \qquad (5.16)$$

Using the above relations to find the derivative of the energy function E with respect to V_i we have

$$\frac{\partial E}{\partial V_i} = \frac{\partial}{\partial V_i} \left[\sum_{i=1}^{n} c_i V_i \right] + \frac{\partial P}{\partial V_i} \qquad (5.17)$$

which in turn yields

$$\frac{\partial E}{\partial V_i} = \sum_{j=1}^{n} c_{ij} + \sum_{j=1}^{m} a_{ji} x_j \qquad (5.18)$$

Also, if E is the Energy Function, it must satisfy the following condition [5]:

$$\frac{\partial E}{\partial V_i} = K C_i \frac{d u_i}{dt} \qquad (5.19)$$

where K is a constant of proportionality and has the dimensions of resistance. Using (5.9), (5.10) and (5.18) in (5.19) results in (for the ith neuron)

$$R_{c1i} = K/a_{1i}$$
$$R_{c2i} = K/a_{2i}$$
$$\vdots \qquad \vdots$$
$$R_{cmi} = K/a_{mi} \qquad (5.20)$$

A similar comparison of the remaining partial fractions for the remaining neurons yields the following:

$$\begin{bmatrix} R_{c11} & R_{c12} & \cdots & R_{c1n} \\ R_{c21} & R_{c22} & \cdots & R_{c2n} \\ \vdots & \vdots & \ddots & \vdots \\ R_{cm1} & R_{cm2} & \cdots & R_{cmn} \end{bmatrix} = \begin{bmatrix} \frac{K}{a_{11}} & \frac{K}{a_{12}} & \cdots & \frac{K}{a_{1n}} \\ \frac{K}{a_{21}} & \frac{K}{a_{22}} & \cdots & \frac{K}{a_{2n}} \\ \vdots & \vdots & \ddots & \vdots \\ \frac{K}{a_{m1}} & \frac{K}{a_{m2}} & \cdots & \frac{K}{a_{mn}} \end{bmatrix} \qquad (5.21)$$

$$\begin{bmatrix} R_1 \\ R_2 \\ \vdots \\ R_m \end{bmatrix} = \begin{bmatrix} K \\ K \\ \vdots \\ K \end{bmatrix} \tag{5.22}$$

The operation of the generalized network of Fig. 5.1 can be understood more clearly by considering some specific examples of small sized problems.

Example 1: One decision variable, no constraints

Consider first the problem of minimization of linear function of a single variable (V_1), as given in (5.23). Also, let us consider that there are no constraints in the minimization problem.

$$f = 2V_1 \tag{5.23}$$

The circuit to minimize (5.23), as obtained from Fig. 5.1 would comprise of a single neuronal amplifier (since only one decision variable is present), and one resistance, R_1 connected at the inverting input terminal of the opamp. Moreover, since there are no constraints to satisfy, the entire synaptic circuitry comprising of the unipolar comparators, and the resistances connected to them, would not be required. The value of the lone resistance R_1 can be obtained from (5.22) by selecting a suitable value of K. If $K = 1 \, K\Omega$, then $R = 1 \, K\Omega$. Applying $c_1 = 2 \, V$, the output of the operational amplifier could be readily obtained as $-V_m$, i.e. the output of the opamp would be saturated at its minimum value, that being decided by the negative biasing of the amplifier. Therefore, it is evident that the task of minimization of (5.23) has indeed been achieved with the value of V_1 reaching the minimum possible voltage at the output of the opamp.

Example 2: One decision variable, one constraint

Next consider that the function to be minimized remains the same as given in (5.23), but now an additional restriction has been imposed on the allowable values of V_1 in the form of a constraint (5.24).

$$V_1 \geq 0 \tag{5.24}$$

Since there is a single constraint, there needs to be one synaptic connection (comprising of the unipolar comparator and resistance). The voltage b_1 equals zero for the comparator. An important point to mention here is the two inputs of the comparator would need to be reversed in comparison with the network of Fig. 5.1, because of the difference in the type of inequality between (5.6) and (5.24). Therefore, the non-inverting input terminal of the comparator would be grounded (as $b_1 = 0$). The inverting terminal would be connected to the output of the neuronal amplifier (to provide V_1 as the input to the comparator). The values of the two resistances, R_{c11} and R_1, can be found from (5.21) and (5.22), as $R_{c11} = 1 \, K\Omega$ and $R_1 = 1 \, K\Omega$.

Now there are two 'forces' acting on the neuronal amplifier. The first is due to the comparator, which outputs a 'high' $(+V_m)$ whenever the voltage V_1 is less than b_1.

The synapse therefore remains active as long as the voltage V_1 remains negative. Consider the reverse situation of V_1 being positive. In that case, the output of the comparator would be $-V_m$, which keeps the diode reverse-biased, thereby effectively cutting off the synapse from the circuit. Thus, it is clear that the non-linear synapse forces V_1 to be non-negative. The other action is due to the presence of the constant voltage c_1 (equal to $2\,V$). This voltage applied to the inverting input terminal of the neuronal amplifier via the resistance R_1 would try to have the output saturated at $-V_m$. The combined effect of the two actions would therefore be to bring the neuron output (V_1) to zero.

It can be verified algebraically that the solution to the problem of minimization of $2V_1$ subject to the constraint that V_1 is non-negative, is at $V_1 = 0$, with the minimum value of the function being zero.

Example 3: Two decision variables, two constraints
Consider the minimization of the function

$$f = 2V_1 + 3V_2 \tag{5.25}$$

subject to

$$\begin{aligned} V_1 &\geq 2 \\ V_2 &\geq 0 \end{aligned} \tag{5.26}$$

Since there are two decision variables, two opamps would be required to emulate the two neurons. Further, since there are two constraints, two non-linear synapses comprising of unipolar comparators would be required. The values of the resistances required can be obtained from (5.21) and (5.22) as

$$R_1 = 1\,K\Omega$$
$$R_2 = 1\,K\Omega$$
$$R_{c11} = 1\,K\Omega$$
$$R_{c12} = \infty$$
$$R_{c21} = \infty$$
$$R_{c22} = 1\,K\Omega$$

and the values of the constant voltages to be applied are

$$c_1 = 2\,V$$
$$c_2 = 3\,V$$
$$b_1 = 2\,V$$
$$b_2 = 0\,V$$

The reader is encouraged to verify that the steady state outputs of the two neuronal amplifiers would be $V_1 = 2\,V$ and $V_2 = 0\,V$, which is also the solution point of the linear programming problem since the function attains a minimum value for these values of the variables (subject to the given constraints).

Lastly, it needs to be mentioned that in all the examples thus far, the coefficients of the decision variables in the constraint inequalities had been deliberately kept as unity. This was done to facilitate an understanding of the operation of the circuit without bothering the reader about the generation of voltages like $2V_1$ and $4V_2$. This could easily be achieved by using a resistive potential divider (after properly scaling the inequalities first), as has been explained in earlier chapters. The same holds true for the realization of constraints like $(2V_1 + 3V_2)$ or $(4V_1 + V_2 + 3V_3)$, which could also be obtained by a suitably designed resistive potential divider.

5.2.1 Hardware Simulation Results

This section deals with the application of the NOSYNN-based network to task of solving various sample LPPs. Firstly, a two variable problem was considered with the objective of minimizing the function

$$5V_1 + 2V_2 \tag{5.27}$$

subject to

$$-V_1 - V_2 \leq -1$$
$$V_2 \leq 1$$
$$V_1 - V_2 \leq 2 \tag{5.28}$$

The values of resistances acting as the weights on the neurons are obtained from (5.21, 5.22). For the purpose of simulation, the value of K was chosen to be $1\,K\Omega$. Using $K = 1\,K\Omega$ in (5.21, 5.22) gives

$$\begin{bmatrix} R_{c11} & R_{c12} \\ R_{c21} & R_{c22} \\ R_{c31} & R_{c32} \end{bmatrix} = \begin{bmatrix} -1\,K\Omega & -1\,K\Omega \\ \infty & 1\,K\Omega \\ 1\,K\Omega & -1\,K\Omega \end{bmatrix} \tag{5.29}$$

$$\begin{bmatrix} R_1 \\ R_2 \end{bmatrix} = \begin{bmatrix} 1\,K\Omega \\ 1\,K\Omega \end{bmatrix} \tag{5.30}$$

For the purpose of PSPICE simulations, the unipolar voltage comparator was realized using a diode clamp with an opamp based comparator. The transfer characteristics obtained during the PSPICE simulations for opamp based bipolar and unipolar comparators are presented in Fig. 5.4. For the purpose of this simulation, the LMC7101A CMOS opamp model from the Orcad library in PSPICE was utilised. The value of β for this opamp was measured to be 1.1×10^4 using PSPICE simulation. The negative values of the resistances as appearing in (5.29) were realized by incorporating additional voltage inverting amplifiers after the diodes.

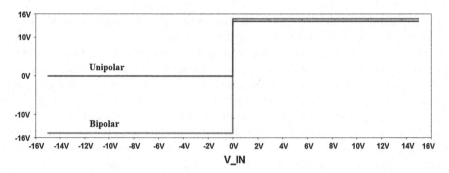

Fig. 5.4 Transfer characteristics for opamp based unipolar and bipolar comparators as obtained from PSPICE simulations

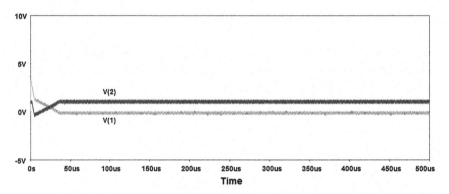

Fig. 5.5 Simulation results for the NOSYNN-based circuit applied to minimize (5.27) subject to (5.28)

Routine mathematical analysis of (5.27) yields: $V_1 = 0$ and $V_2 = 1$. The resultant plots of the neuron output voltages as obtained after PSPICE simulation are presented in Fig. 5.5 from where it can be seen that $V(1) = 141 \, \mu V$ and $V(2) = 1.00 \, V$, which are very near to the algebraic solution thereby confirming the validity of the approach. For emulating a more realistic 'power-up' scenario, random initial values in the milli-volt range were assigned to node voltages. One set of initial node voltage was: $V(1) = 2 \, mV$ and $V(2) = 7 \, mV$.

Next, the NOSYNN-based LPP solver of Fig. 5.1 was employed to minimize the following objective function in three variables

$$4V_1 + 8V_2 + 3V_3 \tag{5.31}$$

subject to

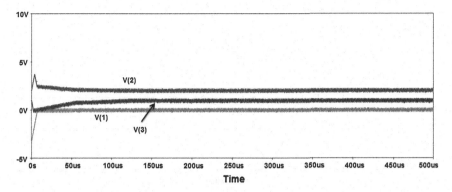

Fig. 5.6 Simulation results for the NOSYNN-based circuit applied to minimize (5.31) subject to (5.32)

$$-V_1 - V_2 \leq -2$$
$$-V_2 - V_3 \leq -5$$
$$-V_1 \leq 0$$
$$-V_2 \leq 0$$
$$-V_3 \leq 0 \qquad (5.32)$$

The values of resistances acting as the weights on the neurons are obtained from (5.21, 5.22). For the purpose of simulation, the value of K was chosen to be $1\,K\Omega$. Using $K = 1\,K\Omega$ in (5.21, 5.22) gives

$$\begin{bmatrix} R_{c11} & R_{c12} & R_{c13} \\ R_{c21} & R_{c22} & R_{c23} \\ R_{c31} & R_{c32} & R_{c33} \\ R_{c41} & R_{c42} & R_{c43} \\ R_{c51} & R_{c52} & R_{c53} \end{bmatrix} = \begin{bmatrix} -1\,K\Omega & -1\,K\Omega & \infty \\ \infty & -0.5\,K\Omega & -1\,K\Omega \\ -1\,K\Omega & \infty & -\infty \\ \infty & -1\,K\Omega & \infty \\ \infty & \infty & -1\,K\Omega \end{bmatrix} \qquad (5.33)$$

$$\begin{bmatrix} R_1 \\ R_2 \\ R_3 \end{bmatrix} = \begin{bmatrix} 1\,K\Omega \\ 1\,K\Omega \\ 1\,K\Omega \end{bmatrix} \qquad (5.34)$$

Routine mathematical analysis of (5.31) yields: $V_1 = 0$, $V_2 = 2$, and $V_3 = 1$. The resultant plots of the neuron output voltages as obtained after PSPICE simulation are presented in Fig. 5.6 from where it can be seen that V(1) = 122 μV, V(2) = 2.01 V and V(3) = 1.00 V, which match closely with the algebraic solutions.

Lastly, a four variable function was considered. The objective was to minimize

$$V_1 + 2V_2 - V_3 + 3V_4 \qquad (5.35)$$

subject to

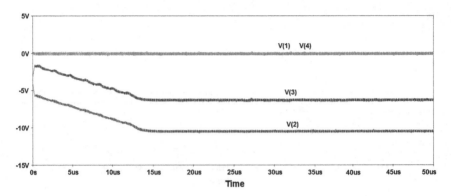

Fig. 5.7 Simulation results for the NOSYNN-based circuit applied to minimize (5.35) subject to (5.36)

$$V_1 - V_2 + V_3 \leq 4$$
$$V_1 + V_2 + 2V_4 \leq 6$$
$$V_2 - 2V_3 + V_4 \leq 2$$
$$-V_1 + 2V_2 + V_3 \leq 2$$
$$V_1 \geq 0$$
$$V_4 \geq 0 \qquad\qquad (5.36)$$

The values of resistances acting as the weights on the neurons are obtained from (5.21, 5.22). For the purpose of simulation, the value of K was chosen to be $1\,K\Omega$. Using $K = 1\,K\Omega$ in (5.21, 5.22) gives

$$\begin{bmatrix} R_{c11} & R_{c12} & R_{c13} & R_{c14} \\ R_{c21} & R_{c22} & R_{c23} & R_{c24} \\ R_{c31} & R_{c32} & R_{c33} & R_{c34} \\ R_{c41} & R_{c42} & R_{c43} & R_{c44} \\ R_{c51} & R_{c52} & R_{c53} & R_{c54} \\ R_{c61} & R_{c62} & R_{c63} & R_{c64} \end{bmatrix} = \begin{bmatrix} 1\,K\Omega & 1\,K\Omega & 1\,K\Omega & \infty \\ 1\,K\Omega & 1\,K\Omega & \infty & 0.5\,K\Omega \\ \infty & 1\,K\Omega & 0.5\,K\Omega & 1\,K\Omega \\ 1\,K\Omega & 0.5\,K\Omega & 1\,K\Omega & \infty \\ 1\,K\Omega & \infty & \infty & \infty \\ \infty & \infty & \infty & 1\,K\Omega \end{bmatrix} \qquad (5.37)$$

$$\begin{bmatrix} R_1 \\ R_2 \\ R_3 \\ R_4 \end{bmatrix} = \begin{bmatrix} 1\,K\Omega \\ 1\,K\Omega \\ 1\,K\Omega \\ 1\,K\Omega \end{bmatrix} \qquad (5.38)$$

Routine mathematical analysis of (5.35) yields: $V_1 = 0$, $V_2 = -10$, $V_3 = -6$ and $V_4 = 0$. The resultant plots of the neuron output voltages as obtained after PSPICE simulation are presented in Fig. 5.7 from where it can be seen that $V(1) = 110\,\mu V$, $V(2) = -10.28\,V$, $V(3) = -6.17\,V$ and $V(4) = 110\,\mu V$ which are very near to the algebraic solution thereby confirming the validity of the approach.

5.3 Non-Linear Feedback Neural Network for Solving QPP

Let the second-order function to be minimized be

$$F = \begin{bmatrix} V_1 \\ V_2 \\ \vdots \\ V_n \end{bmatrix}^T \begin{bmatrix} c_{11} & c_{12} & \cdots & c_{1n} \\ c_{21} & c_{22} & \cdots & c_{2n} \\ \vdots & \vdots & \ddots & \vdots \\ c_{n1} & c_{n2} & \cdots & c_{nn} \end{bmatrix} \begin{bmatrix} V_1 \\ V_2 \\ \vdots \\ V_n \end{bmatrix} \tag{5.39}$$

subject to the following linear constraints

$$\begin{bmatrix} a_{11} & a_{12} & \cdots & a_{1n} \\ a_{21} & a_{22} & \cdots & a_{2n} \\ \vdots & \vdots & \ddots & \vdots \\ a_{m1} & a_{m2} & \cdots & a_{mn} \end{bmatrix} \begin{bmatrix} V_1 \\ V_2 \\ \vdots \\ V_n \end{bmatrix} \leq \begin{bmatrix} b_1 \\ b_2 \\ \vdots \\ b_m \end{bmatrix} \tag{5.40}$$

where $V_1, V_2, \ldots, V_n$ are the variables, and a_{ij}, c_{ij} and b_i ($i = 1, 2, \ldots$, m; $j = 1, 2,$ $\ldots$, n) are constants. The NOSYNN-based circuit to minimize the quadratic function given in (5.39) in accordance with the constraints of (5.40) is presented in Fig. 5.8. As can be seen from Fig. 5.8, individual equations from the set of equations to be solved are passed through non-linear synapses which are realized using unipolar comparators comprising of operational amplifiers and diodes. R_{p1} and C_{p1} are the input resistance and capacitance of the opamp that is used to emulate the functionality of a neuron. These parasitic components are included to model the dynamic nature of the opamp. The outputs of the comparators are fed to neurons having weighted inputs. The neurons are realized by using opamps and the weights are implemented using resistances. The currents arriving to the neuron from various synapses get added up at the input of the neuron.

Applying node equations for node 'A' in Fig. 5.8, the equation of motion of the ith neuron can be given as

$$C_i \frac{du_i}{dt} = \left[\frac{x_1}{R_{c1i}} + \frac{x_2}{R_{c2i}} + \cdots + \frac{x_n}{R_{cni}} \right]$$
$$+ \left[\frac{V_1}{R_{1i}} + \frac{V_2}{R_{2i}} + \cdots + \frac{V_n}{R_{ni}} \right] - \frac{u_i}{R_i} \tag{5.41}$$

where R_i is the parallel equivalent of all resistances connected at node 'A' in Fig. 5.8 and is given by

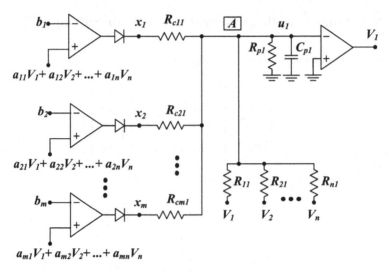

Fig. 5.8 *First* neuron of the NOSYNN-based feedback neural network circuit to solve a quadratic programming problem in n variables with m linear constraints

$$\frac{1}{R_i} = \left[\frac{1}{R_{c1i}} + \frac{1}{R_{c2i}} + \cdots + \frac{1}{R_{cni}} \right]$$
$$+ \left[\frac{1}{R_{1i}} + \frac{1}{R_{2i}} + \cdots + \frac{1}{R_{ni}} \right] + \frac{1}{R_{pi}} \tag{5.42}$$

where u_i is the internal state of the ith neuron, R_{c1i}, R_{c2i}, R_{cni}, ...are the weight resistance connecting the outputs of the unipolar comparators to the input of the ith neuron and R_{1i}, R_{2i}, R_{ni}, ...are the feedback resistances from the outputs of the neurons to the input of the ith neuron. As is shown later in this section, the values of these resistances are governed by the entries in the coefficient matrix of (5.40). R_{pi} and C_{pi} are the input resistance and capacitance of the opamp corresponding to the ith neuron.

Further, as has been explained in Sect. 5.6, the non-linear feedback neural circuit of Fig. 5.8 is associated with the following Lyapunov function (also referred to as the 'energy function')

$$E = \sum_{i=1}^{n} \sum_{j=1}^{n} c_{ij} V_i V_j + \frac{V_m}{2} \sum_{i=1}^{m} \sum_{j=1}^{n} a_{ij} V_j$$
$$+ \frac{V_m}{2\beta} \sum_{i=1}^{m} \ln \cosh \beta \left[\sum_{j=1}^{n} (a_{ij} V_j - b_i) \right] \tag{5.43}$$

This expression of the Energy Function can be written in a slightly different (but more illuminating) form as

$$E = \sum_{i=1}^{n} \sum_{j=1}^{n} c_{ij} V_i V_j + (P_1 + P_2 + \cdots + P_m) \qquad (5.44)$$

where the first term is the same as the second-order function to be minimized, as given in (5.39), and $P_1, P_2, \ldots, P_m$ are the penalty terms. The ith penalty term can be given as

$$P_i = \frac{V_m}{2} \sum_{j=1}^{n} a_{ij} V_j + \frac{V_m}{2\beta} \ln \cosh \beta \left[\sum_{j=1}^{n} (a_{ij} V_j - b_i) \right] \qquad (5.45)$$

Obtaining a partial differentiation of the combined penalty term, P ($= P_1 + P_2 + \cdots + P_m$) with respect to V_i we have

$$\frac{\partial P}{\partial V_i} = \frac{V_m}{2} \sum_{j=1}^{m} a_{ji} + \frac{V_m}{\beta} \sum_{j=1}^{m} a_{ji} \tanh \beta \left[\sum_{j=1}^{n} (a_{ij} V_j - b_i) \right] \qquad (5.46)$$

which may be simplified to

$$\frac{\partial P}{\partial V_i} = \sum_{j=1}^{m} a_{ji} x_j \qquad (5.47)$$

Using the above relations to find the derivative of the energy function E with respect to V_i we have

$$\frac{\partial E}{\partial V_i} = \frac{\partial}{\partial V_i} \left[\sum_{i=1}^{n} \sum_{j=1}^{n} c_{ij} V_i V_j \right] + \frac{\partial P}{\partial V_i} \qquad (5.48)$$

which in turn yields

$$\frac{\partial P}{\partial V_i} = \sum_{j=1}^{n} c_{ij} + \sum_{j=1}^{m} a_{ji} x_j \qquad (5.49)$$

Also, if E is the Energy Function, it must satisfy the following condition [5]:

$$\frac{\partial E}{\partial V_i} = K C_i \frac{du_i}{dt} \qquad (5.50)$$

where K is a constant of proportionality and has the dimensions of resistance. Equation (5.50) applied to the ith neuron results in

$$R_{c1i} = K/a_{1i}$$
$$R_{c2i} = K/a_{2i}$$
$$\vdots \qquad \vdots$$
$$R_{cmi} = K/a_{mi} \tag{5.51}$$

A similar comparison of the remaining partial fractions for the remaining variables yields the following:

$$
\begin{bmatrix}
R_{c11} & R_{c12} & \cdots & R_{c1n} \\
R_{c21} & R_{c22} & \cdots & R_{c2n} \\
\vdots & \vdots & \ddots & \vdots \\
R_{cm1} & R_{cm2} & \cdots & R_{cmn}
\end{bmatrix}
=
\begin{bmatrix}
\frac{K}{a_{11}} & \frac{K}{a_{12}} & \cdots & \frac{K}{a_{1n}} \\
\frac{K}{a_{21}} & \frac{K}{a_{22}} & \cdots & \frac{K}{a_{2n}} \\
\vdots & \vdots & \ddots & \vdots \\
\frac{K}{a_{m1}} & \frac{K}{a_{m2}} & \cdots & \frac{K}{a_{mn}}
\end{bmatrix}
\tag{5.52}
$$

$$
\begin{bmatrix}
R_{11} & R_{12} & \cdots & R_{1n} \\
R_{21} & R_{22} & \cdots & R_{2n} \\
\vdots & \vdots & \ddots & \vdots \\
R_{m1} & R_{m2} & \cdots & R_{mn}
\end{bmatrix}
=
\begin{bmatrix}
\frac{K}{c_{11}} & \frac{K}{c_{12}} & \cdots & \frac{K}{c_{1n}} \\
\frac{K}{c_{21}} & \frac{K}{c_{22}} & \cdots & \frac{K}{c_{2n}} \\
\vdots & \vdots & \ddots & \vdots \\
\frac{K}{c_{m1}} & \frac{K}{c_{m2}} & \cdots & \frac{K}{c_{mn}}
\end{bmatrix}
\tag{5.53}
$$

5.3.1 Simulation Results

This section deals with the application of the NOSYNN-based network to task of minimizing the objective function

$$3V_1^2 + 4V_1V_2 + 5V_2^2 \tag{5.54}$$

subject to

$$V_1 - V_2 \le -1$$
$$V_1 + V_2 \le 1 \tag{5.55}$$

The values of resistances acting as the weights on the neurons are obtained from (5.52, 5.53). For the purpose of simulation, the value of K was chosen to be $1\,\text{K}\Omega$. Using $K = 1\,\text{K}\Omega$ in (5.52, 5.53) gives
$R_{c11} = R_{c12} = R_{c21} = R_{c22} = K = 1\,\text{K}\Omega$, $R_{11} = 1.66\,\text{K}\Omega$, $R_{21} = 1\,\text{K}\Omega$, $R_{12} = 1.66\,\text{K}\Omega$, $R_{22} = 1\,\text{K}\Omega$

For the purpose of PSPICE simulations, the unipolar voltage comparator was realized using a diode clamp with an opamp based comparator, as was done in Sect. 5.2 for the case of the LPP solver. The transfer characteristics obtained during the PSPICE simulations for opamp based bipolar and unipolar comparators are presented

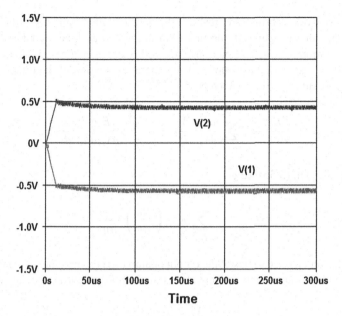

Fig. 5.9 Simulation results for the NOSYNN-based circuit applied to minimize (5.54) subject to (5.55)

in Fig. 5.4. For the purpose of this simulation, the LMC7101A CMOS opamp model from the Orcad library in PSPICE was utilized. The value of β for this opamp was measured to be 1.1×10^4 using SPICE simulation.

Routine mathematical analysis of (5.54) yields: $V_1 = -0.584$, $V_2 = 0.416$. The resultant plots of the neuron output voltages as obtained after PSPICE simulation are presented in Fig. 5.9 from where it can be seen that $V(1) = -0.58$ V and $V(2) = 0.41$ V which are very near to the algebraic solution thereby confirming the validity of the approach. The initial node voltages were kept as $V(1) = -1$ mV and $V(2) = -10$ mV.

5.4 Discussion on Energy Function

This section deals with the explanation of individual terms in the energy function expressions given in (5.12) and (5.43). The last terms in both (5.12) and (5.43) are transcendental in nature and an indicative plot showing the combined effect of the last two terms is presented in Fig. 5.10. As can be seen, one 'side' of the energy landscape is flat whilst the other has a slope directed to bring the system state towards the side of the flat slope. During the actual operation of the NOSYNN-based LPP and QPP solving circuits, the comparators remain effective only when the neuronal output states remain outside the feasible region and during this condition, these unipolar comparators work to bring (and restrict) the neuron output voltages to the feasible region. Once that is achieved, the first term in the energy function takes over and works to minimize the given function. The validity of the energy functions of (5.12) and (5.43) can be proved as follows. The general expression of the time derivative of the energy function 'E' is given by

$$\frac{dE}{dt} = \sum_{i=1}^{N} \frac{\partial E}{\partial V_i} \frac{dV_i}{dt} = \sum_{i=1}^{N} \frac{\partial E}{\partial V_i} \frac{dV_i}{du_i} \frac{du_i}{dt} \qquad (5.56)$$

Using (5.19) in (5.56) we get

$$\frac{dE}{dt} = \sum_{i=1}^{N} K C_i \left(\frac{du_i}{dt} \right)^2 \frac{dV_i}{du_i} \qquad (5.57)$$

The transfer characteristics of the output opamp used to implement the neurons in Figs. 5.1 and 5.8 implements the activation function of the neuron and can be written as

$$V_i = f(u_i) \qquad (5.58)$$

where V_i denotes the output of the opamp and u_i corresponds to the internal state at the inverting terminal. The function f is typically a saturating, monotonically decreasing one, as shown in Fig. 5.11, and therefore,

$$\frac{dV_i}{du_i} \leq 0 \qquad (5.59)$$

thereby resulting in

$$\frac{dE}{dt} \leq 0 \qquad (5.60)$$

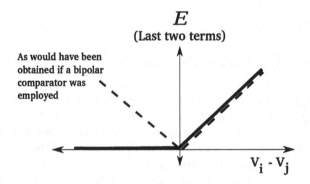

Fig. 5.10 Combined effect of last two terms in (5.12)

with the equality being valid for

$$\frac{du_i}{dt} = 0 \qquad (5.61)$$

Equation (5.60) shows that the energy function can never increase with time which is one of the conditions for a valid energy function. The second criterion *viz.* the energy function must have a lower bound is also satisfied for the circuits of Figs. 5.1 and 5.8 wherein it may be seen that $V_1, V_2, \ldots, V_n$ are all bounded (as they are the outputs of opamps, as given in (5.58) amounting to E, as given in (5.12), having a finite lower bound.

5.5 Issues in Actual Implementation

This section deals with the monolithic implementation issues of the NOSYNN-based circuit. The PSPICE simulations assumed that all operational amplifiers (and diodes) are identical, and therefore, it is required to determine how deviations from this assumption affect the performance of the network. Effects of non-idealities in various components were further investigated in PSPICE by testing the circuit with all resistances having the same percentage deviation from their assigned values. The resulting assessment of the quality of solution for the NOSYNN-based LPP solver is presented in Table 5.1 from where it is evident that the solution point does not change much even for high deviations in the resistance values.

Next, the effect of offset voltages in the opamp-based unipolar comparators was explored. Offset voltages were applied at the inverting terminals of the comparators and the results of PSPICE simulations for the chosen LPP are given in Table 5.2. As can be seen, the offset voltages of the comparators do not affect the obtained solutions to any appreciable extent. However, the error does tend to increase with increasing offset voltages.

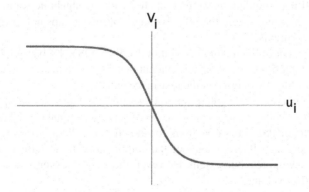

Fig. 5.11 Transfer characteristics of the opamp used to realize the neurons in Fig. 5.1, 5.8

Table 5.1 Effect of variation in resistances on the obtained results of the chosen LPP (5.35)

Percentage variation in resistances (%)	Simulated results (using PSPICE) [V]
+2	$\begin{bmatrix} 108\,\mu V \\ -10.29V \\ -6.17V \\ 115\,\mu V \end{bmatrix}$
+ 5	$\begin{bmatrix} 113\,\mu V \\ -10.35V \\ -6.23V \\ 267\,\mu V \end{bmatrix}$
+10	$\begin{bmatrix} 214\,\mu V \\ -10.77V \\ -6.33V \\ 332\,\mu V \end{bmatrix}$
− 2	$\begin{bmatrix} 98\,\mu V \\ -10.28V \\ -6.17V \\ 129\,\mu V \end{bmatrix}$
− 5	$\begin{bmatrix} 105\,\mu V \\ -9.72V \\ -5.91V \\ 512\,\mu V \end{bmatrix}$
−10	$\begin{bmatrix} 126\,\mu V \\ -9.89V \\ -5.69V \\ 875\,\mu V \end{bmatrix}$

Finally, offset voltages for the opamps emulating the neurons were also considered. Offset voltages were applied at the non-inverting inputs of the opamps and the results of PSPICE simulations were compared with the algebraic solution as given in Table 5.3. As can be seen, the offset voltages of the opamps have little effect on the obtained solutions.

The assessment of the quality of solution for the NOSYNN-based QPP solver is presented in Table 5.4 from where it is evident that the solution point does not change much even for high deviations in the resistance values.

Next, the effect of offset voltages in the opamp-based unipolar comparators was explored. Offset voltages were applied at the inverting terminals of the comparators and the results of PSPICE simulations for the chosen LPP are given in Table 5.5. As can be seen, the offset voltages of the comparators do not affect the obtained solutions to any appreciable extent. However, the error does tend to increase with increasing offset voltages.

Finally, offset voltages for the opamps emulating the neurons were also considered. Offset voltages were applied at the non-inverting inputs of the opamps and the

Table 5.2 Effect of offset voltages of the opamp-based unipolar comparators on the solution quality of the NOSYNN-based circuit on the chosen LPP (5.35)

Offset voltage applied at inverting input of the comparators (mV)	Simulated results (using PSPICE) [V]
−5	$\begin{bmatrix} 103\,\mu V \\ -10.22\,V \\ -6.17\,V \\ 112\,\mu V \end{bmatrix}$
−10	$\begin{bmatrix} 100\,\mu V \\ -10.04\,V \\ -6.11\,V \\ 98\,\mu V \end{bmatrix}$
−15	$\begin{bmatrix} 89\,\mu V \\ -9.92\,V \\ -6.03\,V \\ 77\,\mu V \end{bmatrix}$
+5	$\begin{bmatrix} 127\,\mu V \\ -10.28\,V \\ -6.18\,V \\ 314\,\mu V \end{bmatrix}$
+10	$\begin{bmatrix} 105\,\mu V \\ -10.34\,V \\ -6.29\,V \\ 623\,\mu V \end{bmatrix}$
+15	$\begin{bmatrix} 105\,\mu V \\ -10.84\,V \\ -6.33\,V \\ 746\,\mu V \end{bmatrix}$

results of PSPICE simulations were compared with the algebraic solution as given in Table 5.6. As can be seen, the offset voltages of the opamps have little effect on the obtained solutions.

In fact, the realization of unipolar comparators by the use of opamps and diodes in the NOSYNN-based circuits tends to increase the circuit complexity. The transistor count can be further reduced by utilizing voltage-mode unipolar comparators instead of the opamp-diode combination. This also suggests that a real, large scale implementation for solving linear and quadratic programming problems with high variable counts might be quite different. The resistances used in the synaptic interconnections may be eliminated by employing a voltage comparator with current outputs implemented with a DVCC, as was done in the case of linear equation solver of Chap. 5. Alternative realizations based on the differential equations (5.10) governing the system of neurons are also possible. As an example, to obtain the tanh(.) non-linearity, a MOSFET operated in the sub-threshold region may be employed [6]. Another alternative may be to use a Current Differencing Transconductance Amplifier (CDTA) to provide the same non-linearity in the current-mode regime.

Table 5.3 Effect of offset voltages of the opamp-based neurons on the solution quality of the NOSYNN-based circuit for the chosen LPP (5.35)

Offset voltage applied at the non-inverting input of neuronal opamps (mV)	Simulated results (using PSPICE) [V]
-5	$\begin{bmatrix} 111\,\mu V \\ -10.28\,V \\ -6.17\,V \\ 123\mu V \end{bmatrix}$
-10	$\begin{bmatrix} 119\,\mu V \\ -10.29\,V \\ -6.18\,V \\ 221\,\mu V \end{bmatrix}$
-15	$\begin{bmatrix} 189\,\mu V \\ -10.32\,V \\ -6.20\,V \\ 323\mu V \end{bmatrix}$
$+5$	$\begin{bmatrix} 127\,\mu V \\ -10.12\,V \\ -6.13\,V \\ 314\,\mu V \end{bmatrix}$
$+10$	$\begin{bmatrix} 69\,\mu V \\ -9.98\,V \\ -6.09\,V \\ 623\,\mu V \end{bmatrix}$
$+15$	$\begin{bmatrix} 10\,\mu V \\ -9.89\,V \\ -6.04\,V \\ 746\,\mu V \end{bmatrix}$

5.6 Comparison with Existing Works

For solving a LPP in n variables with m constraints, the NOSYNN based neural network requires n opamps, m unipolar comparators and $(2mn + m + n)$ resistances. In comparison, the network of Rodriguez-Vazquez *et al.* employed n integrators, m summers, a *constraint block* comprising of m comparators and $(m + 1)$ AND gates. The recently NOSYNN-based LPP solver of [7] requires $2n$ resistances, 2 voltage summers, one amplifier and one integrator per variable to solve a given LPP.

For solving a QPP in n variables with m constraints, the NOSYNN based neural network requires n opamps, m unipolar comparators and $(n^2 + 2mn + m)$ resistances. In comparison, the network of Kennedy & Chua comprises of $2n$ opamps, $2n$ resistors and n capacitors for the variable amplifiers whereas for satisfying each constraint, the constraint amplifiers employs $3m$ opamps, $2m$ resistors and m diodes [8]. Wang's network requires $(n + m)$ neurons for solving a QPP in n variables with m constraints. Each neuron is made up of a summer, an integrator, and an

Table 5.4 Effect of variation in resistances on the obtained results of the chosen QPP (5.54)	Percentage variation in resistances (%)	(using PSPICE simulations) [V]
	+2	$\begin{bmatrix} -0.58 \\ 0.41 \end{bmatrix}$
	+5	$\begin{bmatrix} -0.59 \\ 0.43 \end{bmatrix}$
	+10	$\begin{bmatrix} -0.61 \\ 0.44 \end{bmatrix}$
	−2	$\begin{bmatrix} -0.57 \\ 0.41 \end{bmatrix}$
	−5	$\begin{bmatrix} -0.57 \\ 0.40 \end{bmatrix}$
	−10	$\begin{bmatrix} -0.56 \\ 0.38 \end{bmatrix}$

Table 5.5 Effect of offset voltages of the opamp-based unipolar comparators on the solution quality of the NOSYNN-based circuit on the chosen QPP (5.54)	Offset voltage applied at inverting input of the comparators (mV)	(Using PSPICE Simulations) [V]
	−5	$\begin{bmatrix} -0.58 \\ 0.41 \end{bmatrix}$
	−10	$\begin{bmatrix} -0.57 \\ 0.40 \end{bmatrix}$
	−15	$\begin{bmatrix} -0.56 \\ 0.38 \end{bmatrix}$
	+5	$\begin{bmatrix} -0.58 \\ 0.41 \end{bmatrix}$
	+10	$\begin{bmatrix} -0.59 \\ 0.41 \end{bmatrix}$
	+15	$\begin{bmatrix} -0.60 \\ 0.42 \end{bmatrix}$

inverter consuming 3 opamps, 1 capacitor and $(n+5)$ resistors [9]. To solve a QPP in n variables with m constraints, Xia's network consisted of $(2m^2 + 4mn)$ amplifiers, $(2m^2 + 4mn + 3m + 3)$ summers, $(n + m)$ integrators, and n limiters [10]. The recently NOSYNN-based QPP solver of [11] employs $2n$ voltage summers, n integrators, $3n^2$ resistances and $2n$ amplifiers.

5.7 Mixed-Mode Neural Circuits for LPP and QPP

The voltage-mode circuits for LPP and QPP presented in the previous sections employ resistors to set the synaptic weights. In an actual integrated circuit implementation, this would require the fabrication of a large number of resistors thereby consuming

Table 5.6 Effect of offset voltages of the opamp-based neurons on the solution quality of the NOSYNN-based circuit for the chosen QPP (5.54)

Offset voltage applied at the non-inverting input of neuronal opamps (mV)	(Using PSPICE Simulations) [V]
−5	$\begin{bmatrix} -0.58 \\ 0.41 \end{bmatrix}$
−10	$\begin{bmatrix} -0.57 \\ 0.40 \end{bmatrix}$
−15	$\begin{bmatrix} -0.56 \\ 0.39 \end{bmatrix}$
+5	$\begin{bmatrix} -0.58 \\ 0.41 \end{bmatrix}$
+10	$\begin{bmatrix} -0.59 \\ 0.43 \end{bmatrix}$
+15	$\begin{bmatrix} -0.60 \\ 0.44 \end{bmatrix}$

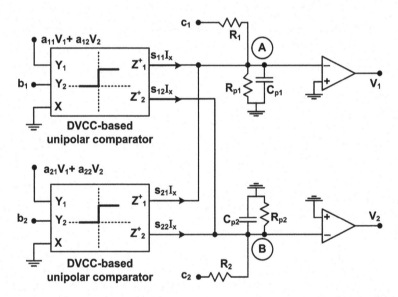

Fig. 5.12 Mixed-mode neural circuit for solving a 2-variable LPP employing DVCC-based unipolar voltage comparators with current outputs

chip area. To obtain a more area-efficient neural circuit, the voltage-mode unipolar comparators (comprising of operational amplifiers and diodes) can be replaced by a unipolar voltage-comparator with current outputs (realized with multi-output DVCC and diodes). A similar technique was employed in the case of the mixed-mode linear equations solver of Chap. 4 wherein the neuron states were voltages whereas currents were used for the synaptic signals.

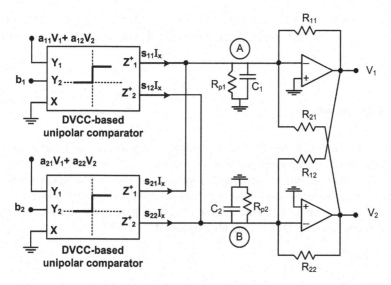

Fig. 5.13 Mixed-mode neural circuit for solving a 2-variable QPP employing DVCC-based unipolar voltage comparators with current outputs

Figure 5.12 presents the mixed-mode neural circuit for solving a 2-variable LPP employing DVCC-based unipolar voltage comparators with current outputs. It is to be noted that the diodes are not shown explicitly. In fact, the DVCC when used as a voltage comparator by directly grounding the X–terminal results in a bipolar comparator with current outputs. To obtain a unipolar comparator, as used in Fig. 5.12, additional diodes need to be connected at the Z+ terminals.

It can be shown that the energy function associated with the network in Fig. 5.12 is of the same form as (5.12) and therefore, the network can be used to solve LPPs in 2-variables. For higher variable counts, the number of neurons would need to be increased (one neuron per variable) in accordance to the general architecture given in Fig. 5.1.

PSPICE simulation of the circuit of Fig. 5.12 employed for solving the 2-variable LPP of (5.27) resulted in $V(1) = 223\,\mu V$ and $V(2) = 1.02\,V$ which are in close agreement with the exact mathematical solution *i.e.* $V_1 = 0$ and $V_2 = 1$. For the PSPICE simulations, DVCCs realized using the CMOS implementation given in Fig. 4.2 and LMC7101A CMOS opamp model from the Orcad library were used. As was done in the case of the voltage-mode circuit for solving LPP, the node voltages were initialized to random values in the milli-volts range for the purpose of PSPICE simulations.

Next, the mixed-mode neural circuit for the solution of QPP in 2 decision variables is presented in Fig. 5.13. As can be observed, the circuit of Fig. 5.13 is a modified form of the network of Fig. 5.8 with the voltage comparators now implemented with DVCCs. PSPICE simulation of the circuit of Fig. 5.13 employed for solving the

2-variable QPP of (5.54) resulted in $V(1) = -0.59\,V$ and $V(2) = 0.41\,V$ which are in close agreement with the exact mathematical solution *i.e.* $V_1 = -0.584$ and $V_2 = 0.416$.

5.8 Conclusion

In this chapter, voltage-mode, CMOS compatible approaches to solve linear and quadratic programming problems in n variables subject to m linear constraints are presented, which employ n neurons and m non-linear synapses. The NOSYNN-based networks were tested on various sample problems and the simulation results confirm the validity of the approach. Further, mixed-mode implementations of the LPP and QPP solvers, based on DVCCs, are also discussed.

References

1. Kreyszig, E.: Advanced Engineering Mathematics, 8th edn. Wiley, New York (2006)
2. Kambo, N.S.: Mathematical Programming Techniques, revised edn. Affilated East-West Press Pvt Ltd., New Delhi (1991)
3. Luenberger, D.G., Ye, Y.: Linear and Nonlinear Programming, 3rd edn. Springer, Heidelberg (2008)
4. Tank, D., Hopfield, J.: Simple 'neural' optimization networks: an A/D converter, signal decision circuit, and a linear programming circuit. IEEE Trans. Circ. Syst. **33**(5), 533–541 (1986)
5. Rahman, S.A., Jayadeva, B., Dutta Roy, S.C.: Neural network approach to graph colouring. Electron. Lett. **35**(14), 1173–1175 (1999)
6. Newcomb, R.W., Lohn, J.D.: Analog VLSI for neural networks. In: Arbib, M.A. (ed.) The handbook of brain theory and neural networks, pp. 86–90. MIT Press, Cambridge (1998)
7. Zhang, Y., Ma, W., Li, X.-D., Tan, H.-Z., Chen, K.: MATLAB simulink modeling and simulation of LVI-based primal-dual neural network for solving linear and quadratic programs. Neurocomputing **72**, 1679–1687 (2009)
8. Kennedy, M.P., Chua, L.O.: Neural networks for nonlinear programming. IEEE Trans. Circ. Syst. **35**(5), 554–562 (1988)
9. Wang, J.: Recurrent neural network for solving quadratic programming problems with equality constraints. Electron. Lett. **28**(14), 1345–1347 (1992)
10. Xia, Y.: A new neural network for solving linear and quadratic programming problems. IEEE Trans. Neural Netw. **7**(6), 1544–1548 (1996)
11. Liu, Q., Wang, J.: A one-layer recurrent neural network with a discontinuous hard-limiting activation function for quadratic programming. IEEE Trans. Neural Netw. **19**(4), 558–570 (2008)

Chapter 6
OTA-Based Implementations of Mixed-Mode Neural Circuits

6.1 Introduction

The Operational Transconductance Amplifier (OTA) is an amplifier which accepts two input voltages and produces an output current which is proportional to the difference of the two input voltages. Therefore, it can be thought of as a voltage-controlled current-source (VCCS). More appropriately, an OTA is a monolithic, direct-coupled, differential-voltage-controlled current-source (DVCCS) [1–10]. The OTA is similar to a standard operational amplifier in that it has a high impedance differential input stage and that it may be used with negative feedback. The point of dissimilarity between an OTA and an opamp is in their output impedances. While the opamp has ideally zero (and practically, very small) output impedance, the OTA exhibits very high (ideally infinite) output impedance.

Since the inputs are voltages and the output is a current, the OTA is best described in terms of its transconductance gain (g_m), rather than the open-loop voltage gain (A). There is usually an additional input for a current (I_{Bias}) to control the amplifier's transconductance. This highlights another point of difference from the opamp. The open-loop gain of an operational amplifier is usually very high and cannot be changed at the user's discretion. The OTA on the other hand, exhibits an open-loop transconductance gain g_m which can be varied linearly over a wide range by adjusting the values of the bias current. Figure 6.1 shows the symbolic diagram of an OTA with the output current being given by

$$I_{out} = g_m(V_1 - V_2) \tag{6.1}$$

where g_m is the transconductance gain of the OTA, and its value is governed by the biasing current (I_{BIAS}) being fed to the device. The characteristics of an ideal OTA may be summarized as follows:

- Input resistance is infinite, $R_{in} \to \infty$
- Output resistance is infinite, $R_{out} \to \infty$
- Bandwidth is infinite, $\omega_o \to \infty$

M. S. Ansari, *Non-Linear Feedback Neural Networks*,
Studies in Computational Intelligence 508, DOI: 10.1007/978-81-322-1563-9_6,
© Springer India 2014

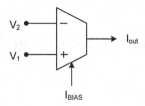

Fig. 6.1 Symbolic diagram of the operational transconductance amplifier (OTA)

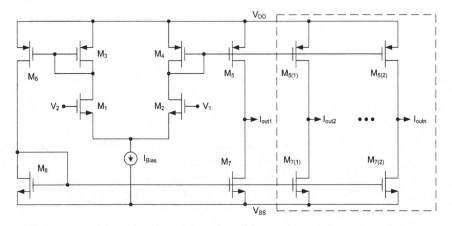

Fig. 6.2 CMOS implementation of the operational transconductance amplifier (OTA) with multiple current outputs

- Perfectly balanced design, $I_{out} = 0$ when $V_1 = V_2$
- Transconductance g_m is finite, and is controllable with the amplifier bias current (I_{BIAS})

The first commercially available integrated circuit units were produced in 1969 by RCA (before being acquired by General Electric), in the form of the hugely popular CA3080, and since then several more integrated circuits have been introduced. With the advent of CMOS technology, OTAs designed and implemented in submicron-CMOS have also been mass produced. One possible CMOS realization of the OTA is presented in Fig. 6.2 where the current I_{Bias} can be used to set the value of g_m. In an actual IC implementation, the current source would actually be realized by a properly biased MOSFET, which makes the CMOS OTA a voltage-controlled current source. Another noteworthy point in Fig. 6.2 is the presence of multiple current outputs. As was done for the case of a DVCC, the output current generating stage of the OTA may be replicated to obtain the desired number of output stages (the extra output stages are shown inside a dotted boundary).

The OTA is not as useful by itself in the vast majority of standard analog signal processing functions as the ordinary op-amp because its output is a current. One of its principal uses is in implementing electronically controlled applications such as variable frequency oscillators and filters and variable gain amplifier stages which are more difficult to implement with standard op-amps. However, for use as a non-linear synapse, a high g_m OTA is very suitable since it can replace the opamp-resistor combination and provide an output current to serve as input to the neuronal amplifier. In essence, the operation of a high gain OTA is very similar to a DVCC with the X terminal grounded. Both of these arrangements are able to realize a voltage comparator with current output.

This chapter presents OTA-based implementations of the various non-linear feedback neural circuits discussed in the previous chapters. The OTA is employed as a voltage comparator with current output(s) and is therefore utilized in the synaptic interconnections between the neurons. Neurons functionality is still being emulated by operational amplifiers. The circuits belong to the so called 'mixed'-mode domain wherein the neuronal states are represented by voltages while the synaptic signals are conveyed as currents. As has been discussed before, this approach alleviates the need of resistances to set the weights thereby reducing the circuit complexity significantly.

6.2 OTA-Based Linear Equation Solver

This section deals with an OTA implementation of the NOSYNN-based linear equation solver discussed in the previous chapters. Readers should recall that a voltage-mode linear equation solver employing non-linear feedback was presented in Chap. 3. The operational amplifier was used both as a voltage comparator as well as an amplifier emulating a neuron. Thereafter, in Chap. 4, it was shown that significant reduction in circuit complexity could be achieved if the opamp-based voltage comparator was replaced by a DVCC-based voltage comparator with scalable current outputs. This possibility of obtaining scaled currents corresponding to voltages fed back from the neuron outputs eliminates the requirement of synaptic resistances.

The OTA of Fig. 6.1 may be made to work as a voltage comparator having a current output, by setting a high value of the transconductance parameter (g_m) through the bias current. In that case, the OTA will behave as a voltage-input current-output saturated amplifier with the output current closely approximated by

$$I_{out} = I_m \tanh \beta(V_1 - V_2) \qquad (6.2)$$

where β is the gain of the amplifier (set to a high value by adjustment of g_m through the bias current) and I_m is the saturation current level of the OTA, and is typically equal to the biasing current (I_{BIAS}).

Let the system of n simultaneous linear equations in n variables be characterized by

Fig. 6.3 NOSYNN-based feedback neural network circuit to solve simultaneous linear equations in n-variables with the synapses implemented using OTAs and opamp emulating the functionality of the neuron

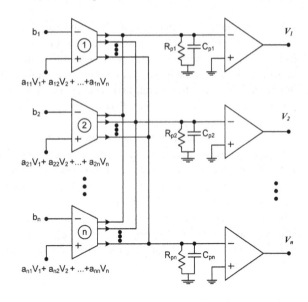

$$\begin{bmatrix} a_{11} & a_{12} & a_{13} & \dots & a_{1n} \\ a_{21} & a_{22} & a_{23} & \dots & a_{2n} \\ a_{31} & a_{32} & a_{33} & \dots & a_{3n} \\ \vdots & \vdots & \vdots & \dots & \vdots \\ a_{n1} & a_{n2} & a_{n3} & \dots & a_{nn} \end{bmatrix} \begin{bmatrix} V_1 \\ V_2 \\ V_3 \\ \vdots \\ V_n \end{bmatrix} = \begin{bmatrix} b_1 \\ b_2 \\ b_3 \\ \vdots \\ b_n \end{bmatrix} \tag{6.3}$$

where V_1, V_2,..., V_n are the variables and a_{ij} and b_i are constants. The NOSYNN-based neural circuit to solve the system of equations of (6.3) is presented in Fig. 6.3, from where it can be seen that, individual equations from the set to be solved are passed through non-linear synapses which are realized using voltage comparators implemented with multiple output operational transconductance amplifiers (MO-OTA). The j-th output of the i-th MO-OTA based voltage comparator can be modelled by

$$I_{out,ji} = s_{ji} I_m \tanh \beta (a_{i1} V_1 + a_{i2} V_1 + \dots + a_{in} V_1 - b_i) \tag{6.4}$$

where V_1, V_2,..., V_n are the neuron outputs, and s_{ij} is the current scaling factor for the j-th output of the i-th OTA. The outputs of the comparators are fed to neurons having weighted inputs. The neurons are realized by using opamps where the weighted (scaled) currents arriving at the input of a neuron from various synapses are added. R_{pi} and C_{p1} are the input parasitic resistance and capacitance of the opamp corresponding to the i-th neuron. These parasitic components are included to model the dynamic nature of the opamp. Node equation for the internal node of the i-th neuron gives the equation of motion as

$$C_{pi} \frac{du_i}{dt} = s_{1i} \, g V_m \tanh \beta (a_{11} V_1 + a_{12} V_2 + \cdots + a_{1n} V_n - b_1)$$

$$+ s_{2i} \, g V_m \tanh \beta (a_{21} V_1 + a_{22} V_2 + \cdots + a_{2n} V_n - b_2) + \cdots +$$

$$+ s_{ni} \, g V_m \tanh \beta (a_{n1} V_1 + a_{n2} V_2 + \cdots + a_{nn} V_n - b_n) - \frac{u_i}{R_{pi}} \quad (6.5)$$

Starting from (6.5), a suitable energy function for the network in Fig. 6.3 can be written as

$$E = \frac{V_m}{\beta} \sum_{i=1}^{n} \ln \cosh \beta \left(\sum_{=1}^{n} a_{ij} V_j - b_i \right) - \sum_{i=1}^{n} \frac{1}{R_i} \int_0^{V_i} u_i d V_i \quad (6.6)$$

From (6.6), it follows that

$$\frac{\partial E}{\partial V_i} = V_m a_{1i} \tanh \beta (a_{11} V_1 + a_{12} V_2 + \cdots + a_{1n} V_n - b_1) +$$

$$V_m a_{2i} \tanh \beta (a_{21} V_1 + a_{22} V_2 + \cdots + a_{2n} V_n - b_2) + \cdots +$$

$$V_m a_{ni} \tanh \beta (a_{n1} V_1 + a_{n2} V_2 + \cdots + a_{nn} V_n - b_n) - \frac{u_i}{R_{pi}} \quad (6.7)$$

Also, if 'E' is the Energy Function, it must satisfy the following condition [11].

$$\frac{\partial E}{\partial V_i} = K C_{pi} \frac{du_i}{dt} \quad (6.8)$$

where K is a constant of proportionality having the dimensions of resistance and is normalized to $1/g_m$ for simplicity. Comparing (6.5) and (6.7) according to (6.8) yields

$$\begin{bmatrix} s_{11} & s_{12} & \cdots & s_{1n} \\ s_{21} & s_{22} & \cdots & s_{2n} \\ \vdots & \vdots & \cdots & \vdots \\ s_{n1} & s_{n2} & \cdots & s_{nn} \end{bmatrix} = \begin{bmatrix} a_{11} & a_{12} & \cdots & a_{1n} \\ a_{21} & a_{22} & \cdots & a_{2n} \\ \vdots & \vdots & \cdots & \vdots \\ a_{n1} & a_{n2} & \cdots & a_{nn} \end{bmatrix} \quad (6.9)$$

PSPICE Simulation Results

The OTA based linear equation solver was first tested for solving the following set of two simultaneous linear equations in two variables.

$$\begin{bmatrix} 1 & 2 \\ 3 & 1 \end{bmatrix} \begin{bmatrix} V_1 \\ V_2 \end{bmatrix} = \begin{bmatrix} 6 \\ 4 \end{bmatrix} \quad (6.10)$$

As was discussed in the previous chapter, some additional circuitry is needed to generate the inputs to the non-linear synapses. An analysis, similar to that performed in previous chapters, yields the following values of the resistors at the input of the OTAs.

Table 6.1 PSPICE simulation results for the NOSYNN-based circuit in which the non-linear synapses are implemented with OTAs, applied to solve various sets of linear equations

[A]	[B]	Algebraic solution [V]	Simulated results (using PSPICE) [V]
$\begin{bmatrix} 1 & 2 \\ 2 & 1 \end{bmatrix}$	$\begin{bmatrix} 3.5 \\ 4 \end{bmatrix}$	$\begin{bmatrix} 1.5 \\ 1 \end{bmatrix}$	$\begin{bmatrix} 1.50 \\ 1.01 \end{bmatrix}$
$\begin{bmatrix} 2 & 1 & 1 \\ 3 & 2 & 1 \\ 1 & 1 & 2 \end{bmatrix}$	$\begin{bmatrix} 5 \\ 10 \\ 6 \end{bmatrix}$	$\begin{bmatrix} -0.5 \\ 5.5 \\ 0.5 \end{bmatrix}$	$\begin{bmatrix} -0.49 \\ 5.50 \\ 0.50 \end{bmatrix}$
$\begin{bmatrix} 2 & 3 & 6 & 1 \\ 3 & 2 & 5 & 4 \\ 2 & 4 & 2 & 5 \end{bmatrix}$	$\begin{bmatrix} 48.75 \\ 67.5 \\ 59.25 \end{bmatrix}$	$\begin{bmatrix} 3.5 \\ 1.5 \\ 5 \\ 7.25 \end{bmatrix}$	$\begin{bmatrix} 3.48 \\ 1.49 \\ 4.95 \\ 7.30 \end{bmatrix}$
$\begin{bmatrix} 2 & 3 & 9 & 2 & 5 \\ 2 & 6 & 9 & 9 & 5 \\ 2 & 6 & 2 & 4 & 5 \\ 2 & 4 & 7 & 8 & 3 \\ 5 & 3 & 6 & 3 & 5 \end{bmatrix}$	$\begin{bmatrix} -54.9 \\ -91.0 \\ -54.9 \\ -74.8 \\ -70.0 \end{bmatrix}$	$\begin{bmatrix} -6.7 \\ -4.3 \\ -2.8 \\ -3.3 \\ 0.64 \end{bmatrix}$	$\begin{bmatrix} -6.76 \\ -4.26 \\ -2.79 \\ -3.32 \\ 0.68 \end{bmatrix}$
$\begin{bmatrix} 1 & 2 & 1 & 3 & 4 & 2 & 1 & 1 & 1 & 2 \\ 2 & 1 & 1 & 3 & 1 & 2 & 2 & 1 & 2 & 3 \\ 1 & 1 & 4 & 3 & 3 & 1 & 1 & 4 & 4 & 1 \\ 4 & 2 & 1 & 5 & 3 & 3 & 1 & 1 & 2 & 2 \\ 1 & 1 & 5 & 1 & 2 & 1 & 2 & 2 & 5 & 2 \\ 3 & 3 & 1 & 2 & 1 & 1 & 5 & 2 & 1 & 1 \\ 5 & 5 & 1 & 4 & 1 & 1 & 3 & 4 & 2 & 2 \\ 1 & 1 & 1 & 2 & 2 & 2 & 3 & 3 & 4 & 1 \\ 4 & 2 & 2 & 1 & 3 & 2 & 5 & 4 & 3 & 2 \\ 1 & 2 & 3 & 1 & 2 & 3 & 1 & 2 & 3 & 4 \end{bmatrix}$	$\begin{bmatrix} 10 \\ -11 \\ 10 \\ 2 \\ -7 \\ -9 \\ -8 \\ -3 \\ -3 \\ 7 \end{bmatrix}$	$\begin{bmatrix} -2 \\ 1 \\ 3 \\ -1 \\ 2 \\ 7 \\ -3 \\ 4 \\ -5 \\ -4 \end{bmatrix}$	$\begin{bmatrix} -2.04 \\ 1.01 \\ 2.97 \\ -1.00 \\ 2.02 \\ 6.99 \\ -3.01 \\ 4.01 \\ -5.04 \\ -3.99 \end{bmatrix}$

$$R_{e11} = 2.5\ K\Omega \quad R_{e12} = 5\ K\Omega \quad R_{e13} = 2.5\ K\Omega$$
$$R_{e21} = 1.66\ K\Omega \quad R_{e22} = 5\ K\Omega \quad R_{e23} = 5\ K\Omega$$

Standard TSMC 0.25 μm parameters were used for the purpose of PSPICE simulations. The supply voltages were set to $V_{DD} = 2.5\ V$ and $V_{SS} = -2.5\ V$. The aspect ratios of the NMOS and PMOS transistors used in the CMOS MO-OTA implementation of Fig. 6.2 were taken to be 20 μm/1 μm for transistors M_1 through M_6 and 5μm/1μm for transistors M_7 & M_8. Further, to get the output currents according to (6.9), currents from individual output stages were added together to provide the required current scaling. For instance, to get a 3 × current scaling, three output stages were combined together yielding an output current equal to $3I_{out}$.

Routine mathematical analysis of (6.10) gives the solution as $V_1 = 0.4$ and $V_2 = 2.8$. The results of PSPICE simulation are found to be $V_1 = 0.389\ V$ and $V_2 = 2.807\ V$ which match closely with the algebraic solution. During the computer simulations, initial conditions in the range of millivolts were imposed on some nodes to model the real-life 'power-up' noise in actual circuits.

Further, more sets of linear equation in 6.3–6.5 and 6.10 variables were solved using the OTA-based non-linear feedback neural circuit. Results for all the cases were found to be in excellent agreement with the algebraic solutions. Table 6.1 contains the results of PSPICE simulations for the chosen sets of equations of varying sizes.

It needs to be mentioned that the technique of combining individual output stages to get the required current scaling is the same as altering the aspect ratios of the transistors in the output stage. Both of these methods result in a circuit which has fixed scaling factors which cannot be changed. To impart reconfigurability to the network, the output stage of the OTA may be replaced by a current summing network (CSN), as was done for the DVCC in the previous chapter, to obtain a digitally controllable OTA. Such a DC-OTA could then be employed in the generalized network of Fig. 6.3 to design a mixed-mode linear equation solver with programmable synaptic weights.

Another important issue to consider is the generation of equal bias currents for the various OTAs used in the network of Fig. 6.3 to ensure matching values of g_m. Generation of replicas of a standard current can be done in many different ways. Figure 6.4 shows one such arrangement in which multiple copies of the reference current can be easily obtained.

6.3 Improved OTA-Based Linear Equation Solver

The mixed-mode linear equation solver of Fig. 6.3 does not require resistances for synaptic interconnections. This makes its circuit complexity favourable as compared to the voltage-mode network presented in Chap. 3. However, some resistances are still required for the potential divider to generate the input voltages for the various comparators. This is primarily because the neuron outputs are voltages which cannot be added like currents (by connecting all currents at one node). Therefore, it is evident that the hardware complexity could be reduced further if the opamps emulating the neurons are replaced by multiple-output OTAs. Such a network in which OTAs are utilized both for realizing the synapses as well as for emulating the neurons, is presented in Fig. 6.5. A grounded resistance (R) would also be required at each of the non-inverting inputs of the OTAs (where the currents fed back from the neurons are added), to convert the summed current to an equivalent voltage. For the sake of simplicity, R may be generalized to unity.

The operation of the network of Fig. 6.5 remains similar to the one presented in the previous section. Individual equations from the set of equations to be solved are passed through non-linear synapses which are realized using voltage comparators (marked as ($C_1, C_2, ..., C_n$) implemented with MO-OTAs. The only difference in the present case is that the OTAs emulating the neurons (marked as $N_1, N_2, ..., N_n$) are also required to have multiple current outputs which can be properly scaled and fed back to the inputs of the comparators to generate the required voltages. A dedicated output current terminal is also shown on each neuron OTA to read the unscaled steady state value of the current. In a set-up where voltage-mode answers are required, these

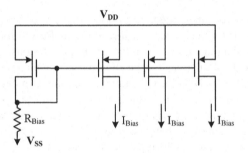

Fig. 6.4 A scheme for generation of equal valued bias currents

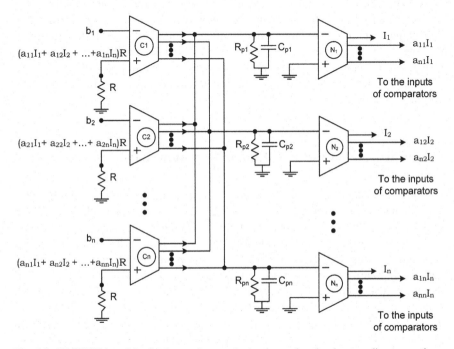

Fig. 6.5 NOSYNN-based feedback neural network circuit to solve simultaneous linear equations in n-variables with the synapses and neurons implemented using OTAs

unscaled current outputs can be readily converted into equivalent voltages by sinking the currents to a resistive load.

Results of PSPICE simulations for the network of Fig. 6.5 for different problems are presented in Table 6.2 from where it can be seen that the steady state point is in close proximity of the algebraic solution in all the test cases. Further, the network converges to the solution point in all the cases. An example depicting the scaling of the currents for a two variable circuit is discussed next. Consider the following system of linear equations:

Table 6.2 PSPICE simulation results for the NOSYNN-based circuit in which the non-linear synapses as well as the neuronal amplifiers are implemented with OTAs, applied to solve various sets of linear equations

[A]	[B]	Algebraic solution [V]	Simulated results (using PSPICE) [V]
$\begin{bmatrix} 1 & 2 \\ 2 & 1 \end{bmatrix}$	$\begin{bmatrix} 3.5 \\ 4 \end{bmatrix}$	$\begin{bmatrix} 1.5 \\ 1 \end{bmatrix}$	$\begin{bmatrix} 1.52 \\ 1.05 \end{bmatrix}$
$\begin{bmatrix} 2 & 1 & 1 \\ 3 & 2 & 1 \\ 1 & 1 & 2 \end{bmatrix}$	$\begin{bmatrix} 5 \\ 10 \\ 6 \end{bmatrix}$	$\begin{bmatrix} -0.5 \\ 5.5 \\ 0.5 \end{bmatrix}$	$\begin{bmatrix} -0.45 \\ 5.54 \\ 0.52 \end{bmatrix}$
$\begin{bmatrix} 2 & 3 & 6 & 1 \\ 3 & 2 & 5 & 4 \\ 2 & 4 & 2 & 5 \end{bmatrix}$	$\begin{bmatrix} 48.75 \\ 67.5 \\ 54 \\ 59.25 \end{bmatrix}$	$\begin{bmatrix} 3.5 \\ 1.5 \\ 5 \\ 7.25 \end{bmatrix}$	$\begin{bmatrix} 3.54 \\ 1.51 \\ 5.03 \\ 7.15 \end{bmatrix}$
$\begin{bmatrix} 2 & 3 & 9 & 2 & 5 \\ 2 & 6 & 9 & 9 & 5 \\ 2 & 6 & 2 & 4 & 5 \\ 2 & 4 & 7 & 8 & 3 \\ 5 & 3 & 6 & 3 & 5 \end{bmatrix}$	$\begin{bmatrix} -54.9 \\ -91.0 \\ -54.9 \\ -74.8 \\ -70.0 \end{bmatrix}$	$\begin{bmatrix} -6.7 \\ -4.3 \\ -2.8 \\ -3.3 \\ 0.64 \end{bmatrix}$	$\begin{bmatrix} -6.81 \\ -4.32 \\ -2.84 \\ -3.31 \\ 0.62 \end{bmatrix}$
$\begin{bmatrix} 1 & 2 & 1 & 3 & 4 & 2 & 1 & 1 & 1 & 2 \\ 2 & 1 & 1 & 3 & 1 & 2 & 2 & 1 & 2 & 3 \\ 1 & 1 & 4 & 3 & 3 & 1 & 1 & 4 & 4 & 1 \\ 4 & 2 & 1 & 5 & 3 & 3 & 1 & 1 & 2 & 2 \\ 1 & 1 & 5 & 1 & 2 & 1 & 2 & 2 & 5 & 2 \\ 3 & 3 & 1 & 2 & 1 & 1 & 5 & 2 & 1 & 1 \\ 5 & 5 & 1 & 4 & 1 & 1 & 3 & 4 & 2 & 2 \\ 1 & 1 & 1 & 2 & 2 & 2 & 3 & 3 & 4 & 1 \\ 4 & 2 & 2 & 1 & 3 & 2 & 5 & 4 & 3 & 2 \\ 1 & 2 & 3 & 1 & 2 & 3 & 1 & 2 & 3 & 4 \end{bmatrix}$	$\begin{bmatrix} 10 \\ -11 \\ 10 \\ 2 \\ -7 \\ -9 \\ -8 \\ -3 \\ -3 \\ 7 \end{bmatrix}$	$\begin{bmatrix} -2 \\ 1 \\ 3 \\ -1 \\ 2 \\ 7 \\ -3 \\ 4 \\ -5 \\ -4 \end{bmatrix}$	$\begin{bmatrix} -2.04 \\ 1.02 \\ 3.02 \\ -0.97 \\ 1.98 \\ 7.05 \\ -3.02 \\ 4.04 \\ -5.02 \\ -4.05 \end{bmatrix}$

$$\begin{bmatrix} 1 & 2 \\ 3 & 1 \end{bmatrix} \begin{bmatrix} V_1 \\ V_2 \end{bmatrix} = \begin{bmatrix} 6 \\ 4 \end{bmatrix} \tag{6.11}$$

The scaling factors for the outputs of the comparators remain the same as found in the previous section. For the first neuronal amplifier, the scaling factors for the two outputs should be $1 \times$ and $3 \times$, and for the second OTA, the scaling factors should be set as $2 \times$ and $1 \times$. These scaling factors could be actually implemented in hardware by combining three similar stages for the second output of the first OTA, and two stages for the first output of the second OTA. For obtaining voltage outputs corresponding to the steady-state currents, additional resistance would be needed at the unscaled outputs of the OTA. Typically, it is advisable to select the value of these resistances as $1/g_m$.

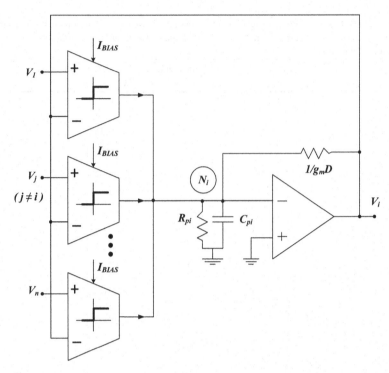

Fig. 6.6 ith neuron of the NOSYNN-based feedback neural network circuit for *graph colouring*

6.4 OTA-Based Graph Colouring Neural Network

In Chap. 2, a NOSYNN-based graph colouring network was presented in Fig. 2.14. The hardware complexity of the circuit could be reduced by utilizing the OTA to implement the voltage comparators needed for the synaptic feedbacks. In that case, the resistances connected at the output of the various comparators would not be needed. Figure 6.6 presents the mixed-mode graph colouring network. As can be seen, the voltage comparators (with current outputs) are realized using single-output OTAs, and the currents coming from the various synapses is added at the input of the neuronal amplifier (implemented using opamp).

The value of the feedback resistance for the neuronal amplifier can be found as $R = 1/g_m D$, where D is the degree of the graph. In a manner similar to the voltage-mode counterpart, feedbacks to the i-th neuron are only from those neurons, which represent nodes to which the i-th node is connected in the graph. Also, it should be noted that the circuit shown in Fig. 6.6 is for one neuron only. For the complete network for colouring a graph comprising of n nodes, n such neurons would be required. For example, to colour a 10 node graph, 10 opamps and at most 100 OTAs would be required.

Table 6.3 PSPICE simulation results for the NOSYNN-based circuit in which the non-linear synapses are implemented with OTAs, applied to solve graph colouring problems

| S. No. | Test Graph | Simulation Results | | Chromatic Number |
		No. of colours	Frequency of occurrence	
1		2	7/10	2
		3	3/10	
2		2	5/10	2
		3	5/10	
3		2	3/10	2
		3	7/10	
4		3	10/10	2

The unipolar comparator characteristics can be obtained by connecting a diode at the output of the OTA, as was done for the case of opamp based comparators in Fig. 2.14. Another option is to bias the OTAs using single-ended power supplies. However, the second approach should be used with caution, as not all available OTA realizations are capable of proper operation using single-ended power supplies.

The network of Fig. 6.6 was tested using computer simulations done on PSPICE software. The results of the simulation tests are presented in Table 6.3, from where it can be seen that the mixed-mode network correctly colours the given graph in all the cases, and the number of colours is also either equal to, or very near to, the chromatic number of the graph. It was also noted that the number of colours assigned by the network was dependent upon the initial condition from which the circuit started. For instance, for the first graph in Table 6.3, the circuit converged to two colours in most of the test cases. However, for some values of initial voltages, the colouring was not minimal, with the network providing three colours as the solution for some trials.

6.5 OTA-Based Neural Network for Ranking

The voltage-mode, non-linear feedback, neural circuit for ranking of numbers discussed in Fig. 2.7 in Chap. 2, can be readily converted into a mixed-mode variant by ensuring the implementation of the voltage comparators by OTAs. By virtue of the use of OTAs, currents are directly available at the output of the comparators and therefore, the syanaptic resistances could be avoided. The resulting mixed-mode network is presented in Fig. 6.7.

The elimination of synaptic resistances reduces the circuit complexity, and takes the network a step closer to an 'all-transistor' implementation. The two remaining resistances (per neuron) could be realized by MOSFET-based circuits. For instance, the grounded resistance can be implemented by the two-transistor circuit given in [12], which is actually a voltage-controlled resistance. The floating resistor, connected as a self-feedback resistance, is more difficult to realize using active devices only [13, 14].

The energy function associated with the network remains of the same form as that of its voltage-mode counterpart of Chap. 2, and is reproduced here, with the appropriate values of resistances put in.

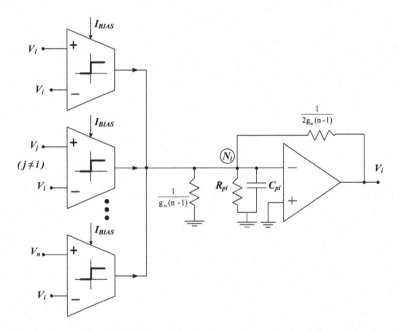

Fig. 6.7 ith neuron of the NOSYNN-based feedback neural network circuit for ranking of numbers

$$E = \sum_{i=1}^{N} \frac{V_i^2}{1/2g_m(n-1)} - \frac{g_m V_m}{2\beta} \sum_{i=1}^{N} \sum_{\substack{j=1 \\ j\neq i}}^{N} \ln \cosh \left(\beta\left(V_i - V_j\right)\right) \qquad (6.12)$$

The first term on the right hand side of (6.12) is quadratic which tries to minimize the number of 'ranks'. The second term has got a negative sign meaning that the energy function E will be minimized if second term is maximized. This happens when the voltages corresponding to different numbers are far away from each other. The first two terms on the right hand side are balancing each other to rank the numbers appropriately.

PSPICE simulation results for the network of Fig. 6.7 yielded promising results. The circuit correctly assigned the highest rank (a voltage equal to $+V_m/2$) to the largest number in the test set, and the lowest rank (a voltage equal to $-V_m/2$) was assigned to the smallest number. Convergence to the correct solution was witnessed in all the test cases, and the time required to rank the numbers was of the order of microseconds.

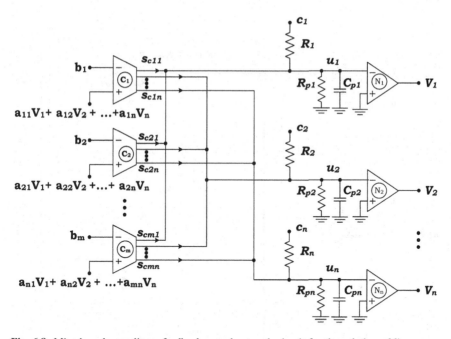

Fig. 6.8 Mixed-mode non-linear feedback neural network circuit for the solution of linear programming problems. The OTAs implement unipolar comparators needed to enforce the constraints

6.6 Linear Programming Using OTAs

The complexity of the voltage-mode LPP solver, presented in Fig. 5.1 in Chap. 5, can be reduced significantly if the resistances connected at the outputs of the unipolar comparators can be avoided. The OTA-based network shown in Fig. 6.8 does exactly that. The voltage comparators are implemented using multiple-output OTAs. The current outputs can be suitably scaled to obtain the required weights for the synaptic interconnections.

As was also discussed for the case of DVCC-based mixed-mode neural circuits in Chap. 4, the required weights can be set at the outputs of the comparators by properly adjusting the aspect ratios of the transistors comprising the output stages in the multiple-output OTAs. The values of the scaling factors can be found using a method similar to the one used in Chap. 4 for DVCC-based comparators.

However, as has been elaborated in previous chapters too, the technique of altering the W/L ratios of the MOSFETs results in a 'hard-wired' circuit suited to solve only one particular LPP. Methods to incorporate programmability in the current values at

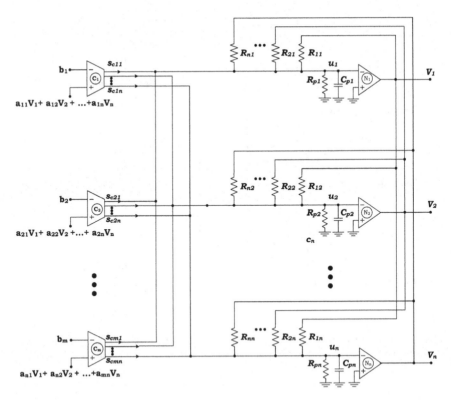

Fig. 6.9 Mixed-mode non-linear feedback neural network circuit for the solution of quadratic programming problems. The OTAs implement unipolar comparators needed to enforce the constraints

the outputs of the OTAs need to be explored. Redesigning the output current stage of the OTA, to include a Current Summing Network (CSN), is one such alternative. Lastly, it needs to be pointed out that resistive potential dividers would still be required at the input of the OTA-based comparators to generate the appropriate input voltage.

6.7 OTA-Based QPP Solver

A mixed-mode, non-linear feedback, neural circuit to solve quadratic programming problems is shown in Fig. 6.9. The multiple output OTAs are used as unipolar voltage comparators with current outputs. The hardware complexity of this mixed-mode network is much lower than that of its voltage-mode counterpart, in which a significant number of passive resistors are required to set the synaptic weights. The resistances connected from the output of neurons to the inputs of other neurons, however, are still present. In a more efficient realization, the neurons themselves may be implemented with multiple-output OTAs. The availability of currents at the outputs of the neurons would result in a resistor-less realization of the QPP solver.

References

1. Parveen, T.: Textbook of Operational Transconductance Amplifier and Analog Integrated Circuits. I.K. International Publishing House Pvt Limited, New Delhi (2009)
2. Zheng, Y.: Operational transconductance amplifiers for gigahertz applications. Canadian Theses, Queen's University (2008)
3. Ferri, G., Sansen, W.: A rail—to—rail constant-g_m low-voltage CMOS operational transconductance amplifier. IEEE J. Solid-State Circuits **32**(10), 1563–1567 (1997)
4. Chatterjeei, S., Tsividis, Y., Kinget, P.: A 0.5-V bulk-input fully differential operational transconductance amplifier. In: Solid-State Circuits Conference, ESSCIRC 2004, Proceeding of the 30th European, pp. 147–150. 2004
5. Yang, S.-H., Kim, K.-Y., Kim, Y.-H., You, Y., Cho, K.-R.: A novel CMOS operational transconductance amplifier based on a mobility compensation technique. IEEE Trans. Circuits Syst. II Express Briefs **52**(1), 37–42 (2005)
6. Chung, W.-S., Kim, K.H., Cha, H.W.: A linear operational transconductance amplifier for instrumentation applications. IEEE Trans. Instrum. Meas. **41**(3), 441–443 (1992)
7. Elwan, H., Gao, W., Sadkowski, R., Ismail, M.: CMOS low-voltage class-AB operational transconductance amplifier. Electron. Lett. **36**(17), 1439–1440 (2000)
8. Wu, P., Schaumann, R.: Tunable operational transconductance amplifier with extremely high linearity over very large input range. Electron. Lett. **27**(14), 1254–1255 (1991)
9. Sánchez-Sinencio, E., Silva-Martinez, J.: CMOS transconductance amplifiers, architectures and active filters: a tutorial. IEE Proc. Circuits Devices Syst. **147**(1), 3–12 (2000)
10. A. Yodtean, A., Thanachayanont, A.: Sub 1-V highly-linear low-power class-AB bulk-driven tunable CMOS transconductor. In: Analog Integrated Circuits and Signal Processing, pp. 1–15, 2013
11. Rahman, S.A., Jayadeva., Dutta Roy, S.C.: Neural network approach to graph colouring. Electron. Lett. **35**(14), 1173–1175 (1999)

12. Ibrahim, M. A., Minaei, S., Kuntman, H.: A 22.5 MHz current-mode KHN-biquad using differential voltage current conveyor and grounded passive elements. AEU - Int. J. Electron. Commun. **59**(5), 311–318 (2005)
13. Singh, S.P., Hansom, J.V., Vlach, J.: A new floating resistor for CMOS technology. IEEE Trans. Circuits Syst. **36**(9), 1217–1220 (1989)
14. Sakurai S., Ismail, M.: A CMOS square-law programmable floating resistor. IEEE Int. Symp. Circuits Syst. ISCAS '93, **2**, 1184–1187 (1993)

Chapter 7
Conclusion

7.1 Conclusion

This book mainly deals with issues in the implementation of neural circuits in actual hardware. A major portion of the text is dedicated to the discussion of NOSYNN-based networks for solving simultaneous linear equations. The feasibility of applying Hopfield network based approaches for the solution of linear equations is first explored wherein it is shown that Hopfield's original network is not suitable for the aforementioned task. Thereafter, modifications that need to be incorporated into the standard Hopfield network in order to obtain a linear equation solver are discussed.

To overcome the limitations of the Hopfield network based linear equation solver, circuits belonging to a class of neural networks called the NOSYNN are presented. By virtue of the use of non-linear feedback, realized using voltage comparators, the energy functions associated with such networks contain transcendental terms thereby making them fundamentally different from the quadratic form of the energy function typically associated with conventional Hopfield networks and its variations.

A voltage-mode realization of NOSYNN based linear equation solver is presented which employs comparators to introduce the required non-linearity in the feedback interconnections between the neurons. Synaptic weights, which are shown to be governed by the coefficients of the linear equations, are implemented using resistances. In a manner similar to the standard Hopfield network, operational amplifiers are utilized for emulating the non-linear characteristics of neurons. For systems of linear equations in two and three variables, the network was tested successfully on a breadboard using standard laboratory components. For large sized problems, results of circuit simulation tests using PSPICE yielded results agreeing with the algebraic solution.

The requirement of resistors for setting the synaptic weights in the voltage-mode linear equation solver can be mitigated by the use of voltage comparators capable of providing scalable current outputs. For this purpose, a relatively new analog building block viz. the digitally-controlled differential voltage current conveyor (DC-DVCC) is discussed, which is then employed to yield a digitally-programmable mixed-mode

M. S. Ansari, *Non-Linear Feedback Neural Networks*,
Studies in Computational Intelligence 508, DOI: 10.1007/978-81-322-1563-9_7,
© Springer India 2014

circuit for solving linear equations. In the DC-DVCC based linear equation solver, neuronal states are voltages whereas the synaptic signals are conveyed as currents. Furthermore, while the internal processing is analog, weights can be set digitally thereby providing flexibility to the circuit. Although the resistors required for synaptic weights can be eliminated by employing DC-DVCC based voltage comparators, the resistances required at the inputs of the voltage comparators are still needed to generate the inputs of the non-linear synapses.

Further, is shown that by incorporating slight modifications in the NOSYNN-based, voltage-mode, linear equation solver, neural networks capable of solving two important mathematical programming problems viz. linear programming problem and quadratic programming problem could be obtained. The energy function associated with both these network is designed in such a way as to first bring the neuronal states to the feasible region and then act to minimize the objective function. Unipolar voltage-comparators (as opposed to bipolar ones used in the linear equation solvers) are needed to obtain the required energy function. Such unipolar comparators were realized by using a diode in conjunction with an operational amplifier. Hardware simulation tests carried out using PSPICE software on a variety of test cases confirmed the validity of the approach with the network converging to the correct solution point in all the test problems. Mixed-mode realizations of the LPP and QPP solvers, based on Differential Voltage Current Conveyors (DVCCs) are also presented wherein it is shown that the hardware complexity of the mixed-mode solvers is much less than their voltage-mode counterparts.

Lastly, a full chapter on Operational Transconductance Amplifier (OTA) based realizations of all the mixed-mode circuits discussed in the book, is included. This is particularly attractive from the viewpoint of setting up test circuits in a laboratory since OTA chips are readily available whereas integrated circuit manufacturers are still reluctant to roll out a DVCC chip.

7.2 Further Reading

In concluding what has been learnt in this book, it is imperative to identify the limitations of the content and what remains to be explored. For the linear equation solving networks discussed in this book, analysis is required to ascertain the response of the networks for the case of linear systems of equations having no solution and infinitely many solutions. Similarly, the behaviour of the networks when the systems of equations is an ill-conditioned one needs to be explored.

Alternative methods to make the weights programmable may be considered. One possible direction of further study is the application of the Current Division Network (CDN) [1–5], to the DVCC to make the output currents digitally controllable. Digital control of Z terminal currents, as obtained using CDN, has been shown to be more versatile than that obtained using the Current Summing Network (CSN) discussed in this book.

In the direction of theoretical analysis, it would be desirable to prove that the stable states of the proposed LPP and QPP networks i.e. the global minima of the energy functions of the corresponding networks, are indeed the solution points of the chosen problems.

Lastly, purely current-mode realizations of all the non-linear feedback neural circuit contained in the book may be explored. A suitable analog building block for starting such an approach is the use of a Current Differencing Transconductance Amplifier (CDTA) [6–10]. The term "current-mode processing" was coined by Barrie Gilbert when he worked on strict translinear loops in which the node voltages are incidental [11]. Current-mode networks are therefore those circuits which exploit current as the main operating parameter i.e. the individual circuit elements should interact by means of currents, and not voltages. Alternatively, current-mode circuits are those where information is represented by the branch currents of the circuits as opposed to nodal voltages in the case os voltage-mode circuits. Such circuits have received considerable research attention owing to the possibility of large dynamic range, wider bandwidth, greater linearity, simple circuitry, low power consumption and less chip area [11, 12]. Another advantage is that the summing of many signals is most readily accomplished when the signals are currents [12]. Therefore, as was to be expected, a lot of research efforts have gone into combining the advantages of ANNs and CM circuits [13–17]. Therefore, an alert reader should strive to come up with CM networks for the various applications discussed in this text.

References

1. Hashiesh, M.A., Mahmoud, S.A., Soliman, A.M.: Digitally controlled cmos balanced output transconductor based on novel current-division network and its applications. In: The 2004 47th Midwest Symposium on Circuits and Systems, MWSCAS'04, vol. 3, pp. 323–326 (2004)
2. Mahmoud, S.A.: Low voltage high current gain cmos digitally controlled fully differential ccii [variable gain amplifier application example]. In: IEEE International Symposium on Circuits and Systems, ISCAS, vol. 2, pp. 1000–1003 (2005)
3. Mahmoud, S.A., Hashiesh, M.A., Soliman, A.M.: Low-voltage digitally controlled fully differential current conveyor. IEEE Trans. Circuits Syst. I Regul. Pap 52(10). 2055–2064 (2005)
4. Ansari, M.S.: Multiphase sinusoidal oscillator with digital control. In: International Conference on Power, Control and Embedded Systems (ICPCES), pp. 1–5 (2010)
5. Tangsrirat, W., Prasertsom, D., Surakampontorn, W.: Low-voltage digitally controlled current differencing buffered amplifier and its application. AEU–Int. J. Electron. Commun. 63(4), 249–258 (2009)
6. Biolek, D.: CDTA—Building block for current-mode analog signal processing. In: Proceedings of ECCTD'03, vol. III, pp. 397–400. Krakow (2003)
7. Biolek, D., Hancioglu, E., Keskin, A.U.: High-performance current differencing transconductance amplifier and its application in precision current-mode rectification. AEU Int. J. Electron. Commun. 62(2), 92–96 (2008)
8. Uygur, A., Kuntman, H.: Seventh-order elliptic video filter with 0.1 dB pass band ripple employing CMOS CDTAs. AEU Int. J. Electron. Commun. 61(5), 320–328 (2007)
9. Siripruchyanun, M., Jaikla, W.: Current-controlled current differencing transconductance amplifier and applications in continuous-time signal processing circuits. Analog Integr. Circ. Sig. Process 61, 247–257 (2009)

10. Biolek, D., Senani, R., Biolkova, V., Kolka, Z.: Active elements for analog signal processing: classification, review, and new proposals. Radioengineering **17**(4), 15–32 (2008)

11. Tomazou, C., Lidgey, F.J., Haigh, D.: Analogue IC design: the current-mode approach. In: IEE Circuits and Systems Series, Institution of Engineering and Technology (IET) (1992)

12. Gilbert, B.: Current mode, voltage mode, or free mode? A few sage suggestions. Analog Integr. Circ. Sig. Process **38**(2), 83–101 (2004)

13. Song, L., Elmasry, M.I., Vannelli, A.: Analog neural network building blocks based on current mode subthreshold operation. In: IEEE International Symposium on Circuits and Systems (ISCAS'93), vol. 4, pp. 2462–2465 (1993)

14. Wu, C.-Y., Lan, J.-F.: CMOS current-mode neural associative memory design with on-chip learning. IEEE Trans. Neural Networks **7**(1), 167–181 (1996)

15. Al-Ruwaihi, K.M.: Current-mode programmable synapse circuits for analogue ulsi neural networks. Int. J. Electron. **86**(2), 189–205 (1999)

16. Balsi, M., Giuliani, G.: Current-mode programmable piecewise-linear neural synapses. Int. J. Circuit Theory Appl. **31**(3), 265–275 (2003)

17. Delgado-Restituto, M., Rodriguez-Vazquez, A.: Current-mode building blocks for CMOS-VLSI design of chaotic neural networks. In: IEEE World Congress on Computational Intelligence, vol. 3, pp. 1993–1997 (1994)

Appendix A
Mixed-Mode Neural Network for Graph Colouring

The modified voltage-mode neural network for graph colouring, presented in Fig. 2.14 [1], was shown to be better than its predecessor of Fig. 2.11 [2]. The primary difference between the two was in the nature of the voltage comparison being done at the input of the synapses. While the original network employed bipolar comparators, the modified graph colouring circuit used unipolar ones, and it was shown in Chap. 2 that the unipolar comparators brought significant performance improvements.

The hardware complexity of the network of Fig. 2.14 can be reduced by replacement of the opamp-based voltage comparators with DVCC-based voltage comparators with current outputs [3]. In that case, the resistances connected at the outputs of the various comparators would not be required. The resulting mixed-mode network is presented in Fig. A.1, from where it can be seen that in order to colour an n-node graph, $n(n-1)$ DVCC-based comparators, n resistances and n operational amplifiers would be required.

In the mixed-mode circuit, output voltages of different neurons represent the colours of different nodes. C_{pi} and R_{pi} denote the internal capacitance and resistance of the ith neuron respectively (included to model the dynamic nature of the opamp), u_i is the internal state and $(1/g_m D)$ is the self-feedback resistance of ith neuron where D is the degree of the graph. The output of other neurons V_j ($j = 1, 2, \ldots, N$) are connected to the input of ith neuron through unipolar comparators which are realized using DVCCs with grounded X-terminals.

As can be seen from Fig. A.1, individual equations from the set of constraints are passed through non-linear synapses which are realized using DVCC-based unipolar comparators. The outputs of the comparators are fed to neurons having weighted inputs. These neurons are realized by using opamps where the currents coming from various comparators act as weights. Weighing is also provided by the self-feedback resistance $(1/g_m D)$. Node equation at the input of the ith neuron gives the equation of motion as

$$C_i \frac{du_i}{dt} = \sum_{\substack{j=1 \\ j \neq i}}^{n} I'_{xj} + g_m D V_i - \frac{u_i}{R_i} \tag{A.1}$$

M. S. Ansari, *Non-Linear Feedback Neural Networks*,
Studies in Computational Intelligence 508, DOI: 10.1007/978-81-322-1563-9,
© Springer India 2014

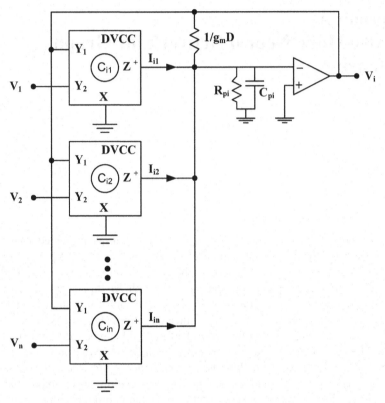

Fig. A.1 ith neuron of the mixed-mode graph colouring neural network based on NOSYNN. DVCCs have been used to realize the voltage comparators with current outputs

where I'_{xj} represents the current output of the jth unipolar comparator, and,

$$\frac{1}{R_i} = g_m D + \frac{1}{R_{pi}} \tag{A.2}$$

The NOSYNN-based graph colouring network of Fig. 2.11, which employs bipolar voltage-mode comparators, is associated with the following energy function [2]:

$$E = \frac{1}{2} \sum_{i=1}^{N} \frac{V_i^2}{2R_{ii}} - \frac{V_m}{4\beta R_c} \sum_{i=1}^{N} \sum_{\substack{j=1 \\ j \neq i}}^{N} g_{ij} \ln \cosh \left(\beta \left(V_j - V_i \right) \right)$$

$$- \sum_{i=1}^{N} \frac{1}{R_i} \int_0^{V_i} u_i dV \tag{A.3}$$

To obtain the energy function for the mixed-mode network of Fig. A.1, we substitute the value of I_{xj} in (A.1) and following the 'intelligent guessing' approach, we get the energy function as

$$E = \frac{g_m D}{2} \sum_{i=1}^{N} \frac{V_i^2}{2} - \frac{g_m V_m}{2\beta} \sum_{i}^{N} = 1 \sum_{\substack{j=1 \\ j \neq i}}^{N} \ln \cosh \left(\beta \left(V_j - V_i \right) \right)$$

$$- \sum_{i=1}^{N} \frac{1}{R_i} \int_{0}^{V_i} u_i dV \qquad (A.4)$$

The last term in (A.4) is usually neglected for high values of the open-loop gain of the opamp used to realize the neurons. The first term on the right hand side of (A.4) is quadratic which tries to minimize the number of colours. The second term has got a negative sign. Therefore, the energy function E will be minimized if second term is maximized. This happens when the voltages corresponding to connected nodes in a graph are far away from each other. The first two terms on the right hand side are balancing each other to colour a graph properly.

The performance of the mixed-mode graph colouring network of Fig. A.1 may further improved by employing unipolar DVCC-based voltage comparators with current outputs. This can be achieved by connecting diodes at the $Z+$ output port of each DVCC. The unipolar characteristics of the comparators would force the network to assign colours from only non-negative values of output voltages, i.e. values between 0 and $+V_m$.

PSPICE Simulation Results

The mixed-mode network, with unipolar comparators, was tested for various random graphs using PSPICE simulations. For the opamp, use was made of the LMC7101A CMOS operational amplifier from National Semiconductor. The sub-circuit file for this opamp is available in Orcad Model Library. Similarly, for the diode the model for D1N3063 diode available in Orcad Library was utilized. The circuit for DVCC was taken from [4]. Standard $0.5\,\mu m$ CMOS parameters were used for simulation purposes. The value of g_m for the DVCC was measured to be 1.091 milli-mhos. The results of the tests are given in Table A.1 from which it is seen that the proposed network gives a solution to all the problems tested and in all the cases the solution is very near to the chromatic number of the graph.

Table A.1 PSPICE simulation results of the proposed circuit applied to color different graphs

S.no	Test graph	Simulation results		Chromatic number
		Number of colours	Frequency of occurrence	
1		2	7/10	2
		3	3/10	
2		2	4/10	2
		3	6/10	
3		3	10/10	2

Appendix B
Mixed-Mode Neural Network for Ranking

A mixed-mode implementation of the NOSYNN-based neural network for ranking of numbers can be obtained by employing an approach similar to the one used in Appendix A. The technique is to replace voltage-mode comparators (realized using opamps) by voltage comparators with current outputs (realized using DVCCs). In such an arrangement, the resistances required to set the synaptic weights are not required thereby reducing the overall hardware complexity of the network. The resulting mixed-mode neural circuit for ranking of numbers is presented in Fig. B.1, from where it is evident that in order to rank n numbers, n opamps, $n(n-1)$ DVCC-based comparators, n grounded resistances and n floating resistances, would be required.

In the mixed-mode circuit, output voltages of different neurons represent ranks of the various input numbers. C_{pi} and R_{pi} denote the internal capacitance and resistance of the ith neuron respectively (included to model the dynamic nature of the opamp), u_i is the internal state and $(1/2g(n-1))$ is the self-feedback resistance of ith neuron where n is the number of neurons. The output of other neurons V_j ($j = 1, 2, \ldots n$) are connected to the input of ith neuron through unipolar comparators which are realized using DVCCs with grounded X-terminals [1].

As can be seen from Fig. B.1, individual equations from the set of constraints are passed through non-linear synapses which are realized using DVCC-based unipolar comparators. The outputs of the comparators are fed to neurons having weighted inputs. These neurons are realized by using opamps where the currents coming from various comparators act as weights. Weighing is also provided by the self-feedback resistance $(1/2g(n-1))$. Node equation at the input of the ith neuron gives the equation of motion as

$$C_i \frac{du_i}{dt} = \sum_{\substack{j=1 \\ j \neq i}}^{n} I'_{xj} + 2g(n-1)V_i - \frac{u_i}{R_i} \qquad (B.1)$$

where I'_{xj} represents the current output of the jth unipolar comparator, and,

M. S. Ansari, *Non-Linear Feedback Neural Networks*, 195
Studies in Computational Intelligence 508, DOI: 10.1007/978-81-322-1563-9,
© Springer India 2014

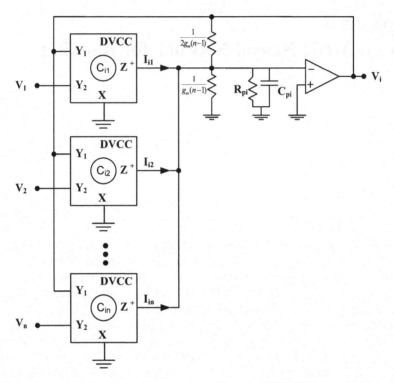

Fig. B.1 ith neuron of the mixed-mode neural network based on NOSYNN for ranking of numbers. DVCCs have been used to realize the voltage comparators with current outputs

$$\frac{1}{R_i} = 2g(n-1) + g(n-1) + \frac{1}{R_{pi}} \tag{B.2}$$

The NOSYNN-based ranking network of Fig. 2.7, which employs bipolar voltage-mode comparators, is associated with the following energy function [2, 3]:

$$E = \sum_{i=1}^{N} \frac{V_i^2}{R/2(n-1)} - \frac{V_m}{2\beta R} \sum_{i=1}^{N} \sum_{\substack{j=1 \\ j \neq i}}^{N} \ln \cosh \left(\beta \left(V_i - V_j \right) \right) \tag{B.3}$$

To obtain the energy function for the mixed-mode network of Fig. B.1, we substitute the value of I_{xj} in (B.1) and following the 'intelligent guessing' approach, we get the energy function as

Table B.1 PSPICE simulation results for the mixed-mode neural network for ranking of two numbers

Numbers to be ranked (Given as initial condition at the neuron outputs) V1, V2 (Volts)	Final steady state neuronal output voltages obtained in PSPICE simulations V1, V2 (Volts)
5, −7	7.391V, −7.391V
0.1, 0.5	−7.391V, 7.391V
2, −3	7.391V, −7.391V
−1, −1.1	7.391V, −7.391V
5, 5.02	−7.391V, 7.391V
0.020, 0.025	−7.391V, 7.391V
0.050, −2	7.391V, −7.391V

$$E = \sum_{i=1}^{n} \frac{V_i^2}{1/2g(n-1)} - \frac{gV_m}{2\beta} \sum_{i=1}^{n} \sum_{\substack{j=1 \\ j \neq i}}^{n} \ln \cosh\left(\beta\left(V_i - V_j\right)\right)$$

$$- \sum_{i=1}^{n} \frac{1}{R_i} \int_{0}^{V_i} u_i dV \qquad (B.4)$$

The last term in (B.4) is usually neglected for high values of the open-loop gain of the opamp used to realize the neurons. The first term on the right hand side of (B.4) is quadratic which tries to minimize the number of ranks. The second term has got a negative sign. Therefore, the energy function E will be minimized if second term is maximized. This happens when the voltages corresponding to the different input numbers are far away from each other. The first two terms on the right hand side are balancing each other to rank the numbers properly.

PSPICE Simulation Results

Verification of the operation of the mixed-mode neural circuit for ranking was carried out using computer simulations done in PSPICE software. It was observed that the network yielded correct results in all the test cases. First, a simple circuit for sorting of two numbers was set up in PSPICE and the network was found to correctly assign a voltage equal to $+V_m/2$ to the larger number and $-V_m/2$ to the smaller number. Some of the test results are presented in Table B.1. As can be seen from the tabulated results, the neural network assigned the higher of the available choices at the output (i.e. $+V_m/2$) to the larger number. The PSPICE simulations were carried out using μA741 operational amplifiers biased at ±15 Volts.

Results of PSPICE simulations of the ranking network configured to sort three numbers are presented in Table B.2. As was done for the case of 2 numbers earlier, the test cases are chosen such that they cater to a wide variety of number types. The network correctly assigned a voltage equal to $+V_m/2$ to the largest number and $-V_m/2$ to the smallest number while assigning a zero (ideally) to the number placed

Table B.2 PSPICE simulation results for the mixed-mode neural network for ranking of three numbers

Numbers to be ranked (Given as initial condition at the neuron outputs) V1, V2, V3 (Volts)	Final steady state neuronal output voltages obtained in PSPICE simulations V1, V2, V3 (Volts)
2, 0, 1	7.391V, −7.391V 232μV
−2,0,1	−7.391V, 232μV, 7.391V
−5, 3, −1.1	−7.391V, 7.391V, 232 μV
0.20, 0.21, 0.15	232 μV, 7.391V, −7.391V
−0.020, −0.017, −0.025	232 μV, 7.391V, −7.391V
7, −7, −5	7.391V, −7.391V 232μV
6.53, −5.90, 1.25	7.391V, −7.391V 232μV
−3, 0, 4.35	−7.391V 232μV, 7.391V
2, 3, 5	−7.391V 232μV, 7.391V

in the middle in the sorting order. The actual voltage, corresponding to the middle number, obtained during the course of PSPICE simulations was not exactly zero and was found to be 232 μV, which can be considered zero for all practical purposes. It is therefore evident that the performance of the ranking network remains the same as that of its voltage-mode counterpart, whereas the complexity of the hardware employed is reduced significantly on account of the removal of the synaptic resistances.

References

1. Mohd. Samar Ansari, Syed Atiqur Rahman, and Syed Javed Arif. Article: A non-linear feedback neural network for graph coloring. International Journal of Computer Applications, 39(16): 31–33: Published by Foundation of Computer Science. USA, New York (February 2012).
2. S.A. Rahman. A nonlinear synapse neural network and its applications. PhD thesis, Department of Electrical Engineering, Indian Institute of Technology, Delhi, India, 2007.
3. Ansari, M.S.: Employing differential voltage current conveyor in graph coloring applications, pp. 1–3. In International Conference on Power, Control and Embedded Systems (ICPCES) (2010).
4. Maheshwari, S.: A canonical voltage-controlled VM-APS with a grounded capacitor. Circuits, Systems, and Signal Processing **27**(1), 123–132 (2008)
5. Jayadeva and S.A. Rahman. A neural network with O(N) neurons for ranking N numbers in O(1/N) time. IEEE Transactions on Circuits and Systems I: Regular Papers, 51(10):2044–2051, 2004.

About the Author

Dr. Mohammad Samar Ansari is an Assistant Professor of the Department of Electronics Engineering at Aligarh Muslim University, Aligarh, India. Before this he worked at the same university as a Lecturer and Guest Faculty from September 2004. Dr. Ansari also worked with Defense Research Development Organization (DRDO) and Siemens Limited during the years 2001–2003. He obtained PhD in 2012 (thesis title: Neural Circuits for Solving Linear Equations with Extensions for Mathematical Programming), and completed MTech (Electronics Engineering) in 2007 and BTech (Electronics Engineering) in 2001 from the same university. He has published 15 international journal papers and more than 30 international and national conference papers. He is a Life Member of The Institution of Electronics and Telecommunication Engineers (IETE), India.

Printed in the United States
By Bookmasters